ARIZONA

Forthcoming Through 1982

Alaska: A Geography, Donald F. Lynch and Roger W. Pearson

Colorado: A Geography, Thomas Melvin Griffiths and Lynnell Rubright

Hawaii: A Geography, Joseph R. Morgan

Maryland: A Geography, James E. Dilisio

Michigan: A Geography, Lawrence M. Sommers

Mississippi: A Geography, Jesse O. McKee

Missouri: A Geography, Milton D. Rafferty

New Mexico: A Geography, Richard W. Helbock and Jerry E. Mueller

North Carolina: A Geography, Ole Gade and H. Daniel Stillwell

South Carolina: A Geography, Charles F. Kovacik and John J. Winberry

Texas: A Geography, Terry G. Jordan, with John L. Bean, Jr., and William M. Holmes

Utah: A Geography, Clifford B. Craig

Wyoming: A Geography, Robert Harold Brown

Geographies of the United States
Ingolf Vogeler, Series Editor

Arizona: A Geography
Malcolm L. Comeaux

This systematic study of the geography of Arizona emphasizes the relationship between the human population and the environment – the patterns of human activities and their effects on the landscape. Dr. Comeaux introduces Arizona's physical features, then traces its history from the time of the early Indians. A discussion of the state's contemporary population and the rapid growth of its cities is followed by a geographic approach to a number of key topics: Arizona's industries – manufacturing, mining, agriculture, lumber, ranching, and tourism – water and land use, and recreation.

Malcolm L. Comeaux, associate professor of geography at Arizona State University, received his Ph.D. from Louisiana State University. He is author of *Atchafalaya Swamp Life: Settlement and Folk Occupations* and has written scholarly articles in a number of journals.

ARIZONA

A GEOGRAPHY

Malcolm L. Comeaux

Routledge
Taylor & Francis Group

LONDON AND NEW YORK

First published 1981 by Westview Press

Published 2018 by Routledge
52 Vanderbilt Avenue, New York, NY 10017
2 Park Square, Milton Park, Abingdon, Oxon OX14 4RN

Routledge is an imprint of the Taylor & Francis Group, an informa business

Library of Congress Cataloging in Publication Data
Comeaux, Malcolm L.
  Arizona: a geography
  (Geographies of the United States)
  Bibliography: p.
  Includes index.
  1. Arizona – Description and travel – 1951–   I. Series.
F811.C73   917.91   80-13119

ISBN 13: 978-0-367-02152-8 (hbk)

ISBN 13: 978-0-367-17139-1 (pbk)

For Michelle and Blaine

CONTENTS

MAPS

FIGURES

INTRODUCTION

The name for the state of Arizona originated from an area south and west of Nogales that the Papago Indians called *Arizonac,* meaning "the place of little springs." A very rich silver strike was made there in 1736, with almost pure silver found on the surface. Many Spaniards moved there, and their mining camp named *Real de Arizona* (the final "c" having been dropped by the Spanish) became well known. The whole area was soon known by the Papago word *Arizona,* and in time the name was given to an entire state to the north (Map 1.1).

Arizona ranks sixth among the 50 states in overall size. It has basically a square shape (Map 1.2), with a maximum north-south distance of 392 mi. (631 km.) and a maximum east-west distance of 338 mi. (544 km.). The total area is 113,956 sq. mi. (295,146 sq. km.). This is a large total area in comparison to many countries, and Arizona is about the same size as Italy, Poland, or the Philippines. But one difference between Arizona and these countries is population. In 1978, Arizona had about 2.5 million citizens, small in comparison to Italy's 50 million and the 35 million of the Philippines.

In theory Arizona has only 22 persons per square mile (8.5 persons/sq. km.), but in reality the people are not evenly distributed across the land. Although its overall population is not dense, Arizona is an urban state, and the vast majority of its citizens live in cities and towns, particularly in the Phoenix and Tucson metropolitan areas. These two areas have great population concentrations, and vast portions of the rest of the state are uninhabited. Most counties have few inhabitants, and one county has well over half the population (Map 1.3). Even in counties with small populations, however, most citizens reside in the many small cities and towns scattered across the land. So although to most outsiders Arizona is a "cowboy and Indian" state,

GENERAL LOCATION

MAP 1.1

actually the vast majority of Arizonans are city dwellers.

Arizona is noted for the many types of environment within its bounds. Many states in the East have a rather uniform environment, but in Arizona the environment ranges from hot, barren deserts receiving less than 5 in. (12.7 cm.) of rain per year to cool forests receiving over 25 in. (63.5 cm.) of precipitation, and even tundra conditions exist at the highest elevations. The key to most of this variation is altitude. In general, the higher the elevation the cooler and damper the climate will be, and in Arizona the altitude varies from about 125 ft. (38 m.) above sea level, along the Colorado River below Yuma, to 12,633 ft. (3,850 m.), at the highest point in the San Francisco Mountains. This total variation of over 12,500 ft. (3,810 m.) fosters many environmental niches for plants and animals.

Arizona is also interesting when examined through time. It has a long history of use by humans beginning with the Indians, continuing through the Spanish and Mexican periods, and concluding in the American era. All these people have left an imprint on the land that can still be seen. Arizona is multi-ethnic today, adding even greater variety to an already colorful state. Indians still occupy and control a large amount of land, and Mexican Amer-

SIZE OF ARIZONA

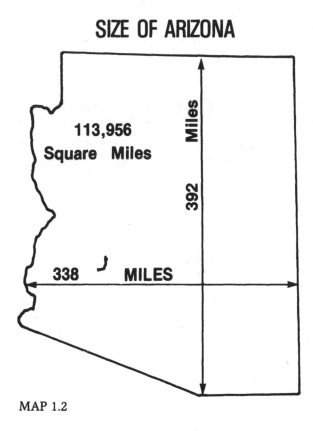

113,956
Square Miles

Miles

392

338   MILES

MAP 1.2

icans form an important minority. Besides these two groups, many other smaller ethnic and racial groups have entered Arizona in the American period.

Arizona can also be characterized as a very dynamic state. Constant changes are occurring in the population and economy. Thousands of people are moving to the state yearly, and only a few are leaving, and, therefore, the trend within recent years has been for population expansion on a large scale. The economy of the state does not hinge on any one sector. Mining, ranching, and agriculture—all traditional industries—remain important; but today manufacturing and tourism are also gaining importance, and this is a healthy trend toward diversification.

The state of Arizona is a dynamic and fascinating place, because of its history and its diversity of environment. Each region, people, city or town, or sector of the economy seems to have its own character and originality. A person does not have to travel far to appreciate something new and different; any short distance traversed with an open mind and an inquiring eye will reveal both. Too many people, tourists as well as Arizonans, consider the long distances found in Arizona an obstacle to be crossed as quickly as possible by freeway, but Arizona is a beautiful state and one that should be enjoyed and appreciated in depth (Map 1.4).

ARIZONA COUNTIES AND POPULATION

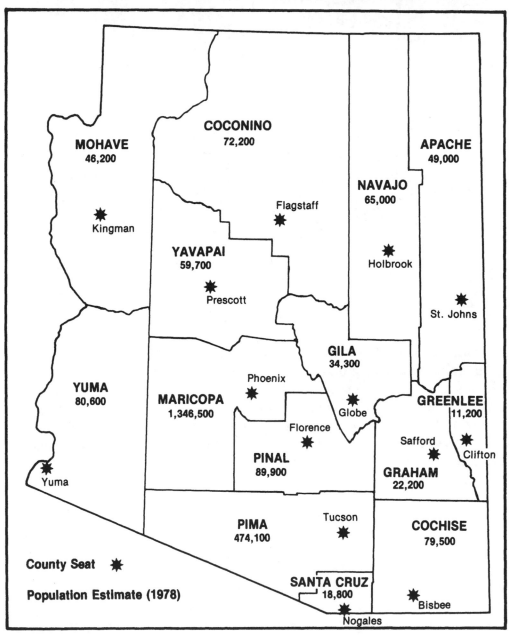

MAP 1.3
Source: Valley National Bank, *Arizona Statistical Review* (1978).

ARIZONA TRANSPORTATION

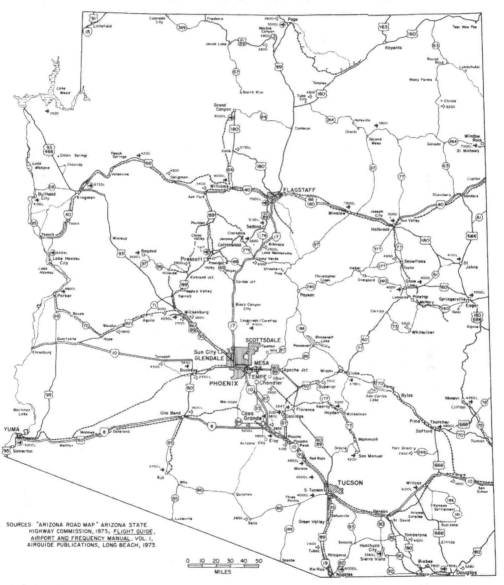

SOURCES: "ARIZONA ROAD MAP" ARIZONA STATE
HIGHWAY COMMISSION, 1973, FLIGHT GUIDE,
AIRPORT AND FREQUENCY MANUAL, VOL. I,
AIRGUIDE PUBLICATIONS, LONG BEACH, 1973.

MAP 1.4
Courtesy: Office of Economic Planning and Development.

2

PHYSICAL BACKGROUND

The range of physical geography found in Arizona is typical of the great contrasts in the state. There are mountains of all ages, shapes, and origins; there are high plateaus, and plateaus upon plateaus; and other areas are noted for broad, open valleys (Map 2.1). This physical variety is accompanied by many climatic variations, with some portions of the state averaging less than 3 in. of precipitation a year, and other regions receiving 10 times that amount. The vegetation also varies greatly, ranging from barren deserts to forests.

An examination of the river patterns clearly illustrates that, in general, Arizona slopes downward from east to west (Map 2.2). Most of the major streams begin in the east and flow toward the west, and the lowest elevations in Arizona are in the far southwestern part of the state. The highest general region is in the central portion of the state along the Mogollon Rim, which forms a divide for rivers flowing north or south. Almost all of Arizona is within the bounds of the Colorado River drainage system with the exception of two areas of internal drainage, Red Lake and Wilcox Dry Lake (both dry lakes), and two small areas that drain southward toward Mexico.

Most of the small streams in the state flowed year-round at the time of European contact; a few, like the Santa Cruz River near Tucson, did tend to stop flowing, but there were always dependable pools of water along their courses. Today most of the small streams are intermittent, flowing only after rainy periods. The reasons why they do not flow year-round vary: some of the streams are dammed; others have large pumps removing the ground-

7

8

LANDFORMS OF ARIZONA

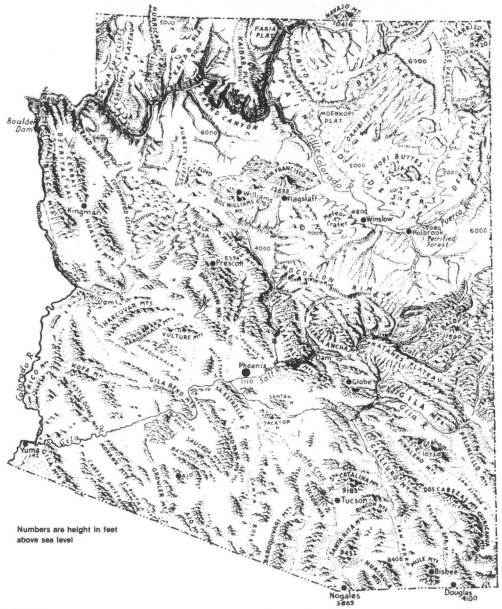

Numbers are height in feet
above sea level

MAP 2.1
Source: "Landforms of the United States," by Erwin Raisz. Courtesy, Mrs. Erwin Raisz.

SCALE: 1 IN ≈ 67 MI

RIVERS AND STREAMS

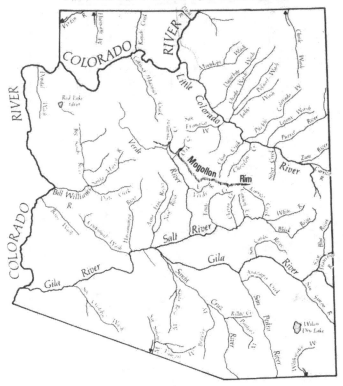

MAP 2.2

water so the surface water tends to soak downward; and drainage areas are adversely affected by overgrazing.

Many newcomers and travelers laugh at western "rivers," for they are often dry and would go unnoticed in the East. In a dry area like Arizona, however, these streams are very important. Water availability determined many events and decisions of the past, and water and past decisions influence Arizona today. All the early routes across the state, both rail and road, carefully considered water availability and followed the streams as much as possible; modern highways still follow the old routes, though water must now be obtained by pumping. Also, water was a consideration when most early towns were settled, and water is as important to these towns today as it was then.

Arizona has few clear boundaries that easily divide the state into definite physical regions. But no one would disagree that the state can be divided into three basic areas—the Plateau to the north, the Basin and Range to the south and west, and the Mountain District in the middle (Map 2.3). Defining the boundaries to these districts, however, does lead to controversy.

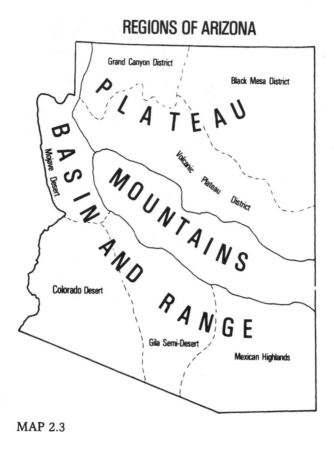

MAP 2.3

PHYSICAL REGIONS

Plateau Region

The Plateau region occupies approximately one-third of the state and is located in the north and northeast. This area is part of a larger region known as the Colorado Plateau, which extends into the neighboring states of Utah, Colorado, and New Mexico. All of the Colorado Plateau is at a high elevation, ranging from 4,300 to 9,000 ft. (1,311 to 2,743 m.), and the only areas below 5,000 ft. (1,524 m.) are the valleys and canyons cut by streams. The entire area has been uplifted, and the sedimentary rocks, mostly Paleozoic and Mesozoic in age, were uplifted evenly with little deformation of the rocks. Most of the rocks of this area are nearly horizontal over large areas (Figure 2.1) though in a few regions they are broken by faults or pushed into monoclines. The rocks of this region are actually of all ages: the basement is Precambrian, above it are the structures of Paleozoic and Mesozoic ages, and these are then capped in many areas with recent

SIMPLIFIED CROSS SECTION OF COLORADO PLATEAU AND GRAND CANYON

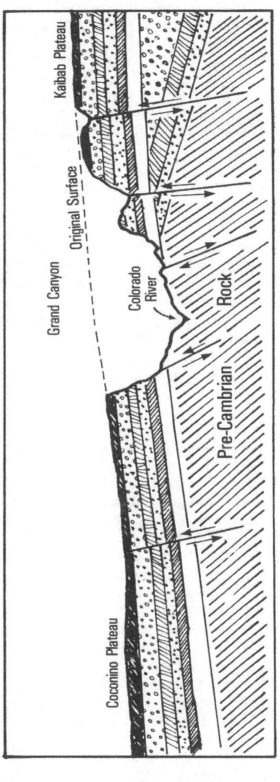

FIGURE 2.1

Cenozoic rock of sedimentary and volcanic origin.

The border of the Plateau region is generally well defined. The most obvious and striking border is to the south along the Mogollon Rim (pronounced mŭg-ē-on′ and named after a former Spanish governor of New Mexico). The Mogollon Rim, often called "the Rim" in Arizona, is a nearly vertical cliff of 1,000 to 2,000 ft. (305 to 610 m.) in height and is one of the most prominent and distinctive features in the state. The Rim is an old fault line; above it are the flat sedimentary rocks of the Plateau, and to the south are the greatly deformed rocks that make up the Mountain District.

The Rim in Arizona begins south and west of Williams and extends east, southeast. Along the western end it follows the upper Verde Valley, and it is well defined in this area except for the southern end of the Verde Valley where the Rim is concealed by a younger lava flow. Almost all precipitation falling on the Plateau near the Rim flows northward, but this western region is an exception with several streams flowing south. These streams have a very steep gradient as they exit the Plateau and have cut colorful and picturesque canyons. The most prominent of these is Oak Creek Canyon, which was once a very important gateway because it was the major route onto the Plateau from the south. With completion of a freeway between Phoenix and Flagstaff, this canyon lost much of its significance, but it remains an important recreational area and is a route used by tourists because of its beauty.

The westernmost edge of the Plateau is also well defined, and the Grand Wash Cliffs, an old fault scarp, form the western boundary. These cliffs are 1,000 to 4,000 ft. (305 to 1,219 m.) in height, and the maximum vertical displacement along this fault is enormous, about 7,000 ft. (2134 m.). The cliffs are formidable and are crossed by no major roads. To the north, in Utah, the Grand Wash Cliffs gradually decline, and there the border of the Plateau shifts to the Hurricane Cliffs. The Hurricane Cliffs also extend into Arizona, but they are of particular importance in Utah where there are several towns, such as Cedar City, at the base of the fault scarp.

Between the Grand Wash Cliffs and the Verde Valley, the precise boundary of the Plateau is open to question. Just south of the Grand Wash Cliffs is an easy gradient that very early became the major gateway to the Plateau from the west. This route, just west of Peach Springs, is used by the Santa Fe Railroad and old U.S. Route 66, though the new Interstate 40 exits the Plateau further south. South of this gateway the Cottonwood Cliffs seem to continue the Grand Wash Cliffs as a southwestern edge of the Plateau. Between the Cottonwood Cliffs and the upper Verde Valley, however, the edge of the Plateau cannot be discerned. In this region the Plateau slowly grades into the Mountain District.

The boundary of the Plateau to the east along the border with New Mexico is also very obscure. The Mogollon Rim does continue through this area, but here it is covered by the geologically recent lava flows that formed the White Mountains (Map 2.1). The exact boundary of the Plateau is almost impossible to find since the Rim is covered by lava and in places it is several thousand feet thick. Past researchers have seldom agreed as to just which physical province the White Mountains belong: some say they are a part of the Mountain District; others say they are a part of the Plateau; still others simply divide the mountain range in half by picking the highest point and saying that everything to the north is a part of the Plateau and that everything to the south is a part of the Mountain District.

This study considers the White Mountains to be a part of the Plateau region for several reasons. The surface forms, geological structure, and history of these mountains are similar to those of the San Francisco Mountains, and no one would consider those mountains separate from the Plateau. Also, the rock beneath the volcanic material is mostly the same as on the Plateau, except for its southernmost edge – flat, horizontal sedimentary rock. The White Mountains are thus considered to be on top of the Plateau and a part of that region. The southern edge of the Plateau region should therefore follow the southern edge of the White Mountains. But where do the White Mountains end? Even that is an unclear boundary.

The Plateau can be divided into three broad structural regions or districts: Grand Canyon, Black Mesa, and Volcanic Plateau (Map 2.3).

Grand Canyon Region. The Grand Canyon region includes the Grand Canyon, the Colorado River as it traverses the Plateau, and all of the Plateau north of this area. The region north of the Colorado River is known to Arizonans as "the Strip" and to outsiders as the "Arizona Strip." The Strip is noted mostly for its sparse settlement and isolation from the rest of the state. It is approximately the size of New Jersey, but there are only two small towns, Fredonia, and the new and rapidly growing town of Colorado City, and only one major highway crosses it. People living in the Strip are isolated from the rest of Arizona: even to go to a county seat requires traveling around the Grand Canyon, either by way of Utah and Nevada or, eastward, crossing the Colorado at Navajo Bridge. Residents of the Strip have particularly close ties with Utah. Most of their services, such as newspapers and groceries, come from Utah, and the vast majority of the residents are Mormons and therefore have emotional ties to the area to the north. On several separate occasions, Utah has even tried to acquire this area from Arizona.

The Strip is noted for several parallel north-south trending faults, and each separates the area into subregions. The western edge of the Plateau, the well-defined Grand Wash Cliffs, follows an old fault line. The Plateau rises vertically at this point and overlooks the Basin and Range Province to the west. Eastward from these cliffs is the first raised section of the Plateau known as the Shivwits Plateau, which is capped with lava in places, as is common in other parts of the Colorado Plateau. Eastward is the second of the north-south trending faults, and it forms the Hurricane Cliffs, which rise step-like another 1,800 ft. (549 m.). East of the Hurricane Cliffs is the Kanab Plateau with Kanab Creek along its eastern edge. Kanab Creek is a small stream that follows another of the north-south fault lines. The stream has cut a very deep canyon near the Colorado River, but close to the Utah border it is at the surface and is Fredonia's source of water.

The Kaibab Plateau, east of Kanab Creek, is the largest and highest of the plateaus in the Strip. The rocks are not completely horizontal in this area, but are warped up into a gentle anticline, so that this plateau averages about 8,000 ft. (2,438 m.) in elevation. This plateau is heavily forested and, in places, averages over 25 in. (63.5 cm.) of rain per year. The Kaibab area is the highest, wettest, and coolest portion of the Strip.

House Rock Valley, a broad, open, and low valley, is east of the Kaibab Plateau, and in the northeasternmost part of the Strip there is another fairly high plateau, the Paria Plateau. The Paria Plateau is not high enough to support forests, so the vegetation is mostly grasses, bushes, and in a few areas, juniper trees. The Paria Plateau is isolated from the rest of Arizona by very high cliffs of a reddish color, the Vermillion Cliffs. These are very formidable, and from the south there are only one or two small Indian trails leading to the top.

The Grand Canyon (Figure 2.2) extends from approximately the mouth of

FIGURE 2.2. The Grand Canyon as seen from the South Rim. Note how the rocks on the north side are in flat layers.

the Little Colorado River westward to the Grand Wash Cliffs, a distance of about 200 mi. (322 km). The Grand Canyon is about 5,000 ft. (1,524 m.) deep and is 5 to 10 mi. wide (8 to 16 km.). The canyon was caused by the cutting action of the Colorado River; as the Plateau was slowly uplifted, the Colorado River cut downward. The whole process began about 10 million years ago, about the middle of the Pleistocene epoch. The river is still cutting downward, but it is now digging into very resistant Precambrian rock and is cutting down at the rate of only about 1 ft (30.5 cm.) each 2,000 years. A trip down into this canyon makes an excellent geologic field trip, as all ages of rock are encountered from the very recent at the surface to ancient Precambrian at the bottom.

Between Utah and the Grand Canyon, the Colorado River travels through Marble Canyon (Figure 2.3), which is now part of the Grand Canyon National Park. Marble Canyon is a very steep-sided and narrow canyon, and from the air it looks like a knife has slashed a narrow cut in a broad open plateau (Figure 2.4). There is only one easy, natural crossing of the Colorado River in Arizona, and that is north of Marble Canyon near the mouth of the Paria River. This crossing, known as Lee's Ferry, was the major route used by the Mormons who came into Arizona. The site is not used today, and the only highway crossing the Colorado River in this part of Arizona is a few miles south of Lee's Ferry at Navajo Bridge.

Black Mesa Region. The Black Mesa region in northeast Arizona is also sometimes known as Navajo country (though the Hopi may disagree with that term) or the Four Corners area. It is, however, the Black Mesa that dominates the region, and the area is generally known by that name. This area is fairly similar to the Grand Canyon region, except that the rocks are of a somewhat younger age and the area lacks the large number of north-south fault lines. It is generally a flat, open desert country that has been greatly affected by erosion (Figure 2.5).

The border on the west is Marble Canyon, and just east of it are the Echo Cliffs. Eastward from the Echo Cliffs is the open and relatively flat Kaibito Plateau. The southwest border of this area is the Painted Desert (Figure 2.6), which borders the Colorado and Little Colorado rivers. There are no set limits to the Painted Desert; it just slowly begins and ends, and depending on light conditions, that beginning and ending will vary. The colors in the landscape, ranging from reds, yellows, oranges, browns, and greens, are the result of various oxidation stages of iron oxides in the rocks.

The southern boundary of the Black Mesa district generally follows the Little Colorado River. The Hopi Buttes in this area are made up of volcanic material such as dikes, plugs, and volcanic flows. The entire area is greatly eroded, and the plugs for former volcanic cones are very prominent features on the landscape. The Petrified Forest is also a part of the southern border. Like the Painted Desert, it just slowly begins and ends, although there are very specific bounds to the Petrified Forest National Park.

FIGURE 2.3. Marble Canyon looking south from Navajo Bridge, a few miles downstream from Lee's Ferry. The Echo Cliffs can be seen in the distance on the left.

FIGURE 2.4. View westward from the top of the Echo Cliffs. In the middle distance can be seen the deep gash that is Marble Canyon. The Vermillion Cliffs and the Paria Plateau are in the background, and the Kaibab Plateau is to the far left.

FIGURE 2.5. The open expansiveness of the Black Mesa area. In the far left background can be seen a Navajo dwelling.

FIGURE 2.6. The Painted Desert. This is a badly eroded area that resembles a "badlands."

The northern border of the Black Mesa area follows the Utah boundary, though a good physical boundary would be the San Juan River in Utah. Two significant landscape features are Navajo Mountain and Monument Valley. Navajo Mountain on the Utah border just east of Page is a volcanic feature. This mountain developed when huge masses of molten rock rose but did not reach the surface, spreading out instead between layers of sedimentary rock. The rocks above the lava domed upward to form the mountain, which rises several thousand feet higher than the plateau around it. The lava is nowhere exposed, so the feature is termed a laccolith rather than a volcano. Monument Valley is also on the Arizona-Utah border (Figure 2.7). This broad, open valley has a number of tall, isolated spires and mesas, which were caused by erosion, or rather a lack of it. As Monument Valley was eroded to its present condition, these tall, isolated buttes managed to resist that erosion.

The very center of this district is the Black Mesa Plateau. It is actually a broad, general syncline with the rocks dipping downward, but it remains a high plateau because the rocks on top have not been eroded. The overlying rocks forming the plateau are fairly young and are rich in coal. The entire plateau is made up of many angular formations including smaller plateaus, broad and narrow valleys, cliffs, and other similar features. The southern

FIGURE 2.7. View in Monument Valley.

end of the Black Mesa Plateau is deeply incised by several washes flowing south toward the Little Colorado. As these washes have cut downward on the Black Mesa, they have left several broad plateaus or mesas, resembling peninsulas, extending southward. These mesas are the homeland of the Hopi Indians, and from east to west are First Mesa, Second Mesa, and Oraibi Mesa (or Third Mesa).

Eastward from the Black Mesa Plateau is Chinle Valley. It is a broad, open valley with several Navajo settlements and some agriculture. Chinle Wash flows northward through this valley toward the San Juan River.

East of Chinle Valley is Defiance Plateau, the last major physical feature in the Black Mesa region. The Defiance Plateau is actually a large anticlinal upland having a plateaulike summit. There is an escarpment on the western edge overlooking Chinle Valley, so to the casual observer the Defiance Plateau does resemble a plateau. There are several deep vertical canyons cut into this region, the most prominent being Canyon de Chelly, now a national monument, which is a very steep-walled canyon (Figure 2.8). Canyon de Chelly was the last retreat of the Navajo, and when Kit Carson entered the canyon and destroyed the crops in the 1860s, the back of the last Navajo rebellion was broken. The canyon thus has a special meaning and signifi-

cance for the Navajo. The southernmost side of Defiance Plateau has a V-shaped valley cut by Black Creek, which flows southward toward the Puerco River. There are several Navajo settlements in this area, including the tribal headquarters at Window Rock.

North of the Defiance Plateau are two mountain groups–the Chuska Mountains, known to the Navajo as Lukachukai, and further north, the Carrizo Mountains. They are both volcanic in origin and are high enough (over 9,000 ft. [2,743 m.]) to receive enough moisture to support ponderosa pine forests.

Volcanic Plateau Region. The Volcanic Plateau is that region of the Colorado Plateau south of the Grand Canyon and the Little Colorado River. There are two major volcanic portions, the White Mountains and the area around the San Francisco Mountains. The White Mountains, a large volcanic mass on the eastern border of the state, are mostly on the Plateau but do spill over onto the Mogollon Rim. The highest point in the White Mountains, Baldy Peak, reaches 11,590 ft. (3,533 m.).

The other prominent volcanic area is the mountain region immediately north of Flagstaff in the portion of the Plateau known as the San Francisco Plateau. The most striking feature of this area is the San Francisco Mountains (Figure 2.9). The tops of these peaks are in a rough horseshoe shape, with a broad valley in the middle descending to the northeast. This is Interior Valley, and around it are six prominent peaks: Humphreys, Agassiz, Fremont, Doyle, Reese, and Abineau. Humphreys Peak is the highest, rising 5,000 ft. (1,524 m.) above the level of the Plateau to a total height of 12,633 ft. (3,850 m.), the highest point in Arizona. The San Francisco Mountains were once probably 15,000 ft. (4,572 m.) high when first formed, but they have since eroded to the present heights. There are other impressive volcanic mountains in this area; the most prominent are to the west and are Bill Williams Mountain (9,264 ft. [2,824 m.]), southwest of Williams, and Sitgreaves Mountain (9,600 ft. [2,926 m.]), and Kendrick Peak (10,418 ft. [3,175 m.]) to the northeast of Williams. Extending northward from the San Francisco Plateau toward the Grand Canyon is the Coconino Plateau. It is a broad, relatively flat area with a cover of volcanic material on the surface. This plateau is divided into two sections by a deep narrow canyon cut by Cataract Creek.

There were three broad periods of eruption in this portion of the Plateau. The first period extended from about 6 to 2 million years ago. In this early phase, about 2,000 sq. mi. (5,180 sq. km.) of land were covered with basalt in sheets of 50 to 300 ft. (15 to 91 m.) thick. Eruptions from the second period are rich in quartz, and these eruptions formed the San Francisco

FIGURE 2.8. Canyon de Chelly. Courtesy: U.S. Department of the Interior, National Park Service photograph by Fred E. Mang, Jr. (Neg. #NPS-5791-55).

FIGURE 2.9. The San Francisco Mountains in early springtime.

Mountains and some other nearby features. The most recent period of eruption, which continued almost to the present (if indeed it has ended), formed the many cinder cones and basalt flows in the area. The last eruption from this period formed Sunset Crater, north and east of the San Francisco Mountains and now a national monument, and its formation has been dated between the autumn of A.D. 1064 and the spring of 1065. Geologically this is very recent, and no one should be surprised if once again volcanic action should occur in this area.

The last section of the Volcanic Plateau is the region between the White Mountains and the San Francisco Mountains known as the Mogollon Plateau. The Mogollon Plateau is a broad, open, and flat area that slopes gently downward from the Mogollon Rim to the Little Colorado River. The highest portions near the Rim are forested, but as one travels northward the vegetation changes to grassland, and near the Little Colorado River it is desert. An interesting feature in this area is Meteor Crater, caused by a nickel-iron meteor striking the earth. There have been attempts to drill straight down the center of the crater to perhaps mine the meteor, but all attempts have met with failure. It is now believed that the meteor struck the earth at an angle from the north, so that the meteor is buried under the south rim of the crater. This has been tentatively confirmed by drilling in that area.

Basin and Range Province

The Basin and Range Province is very large and extends far beyond the bounds of Arizona. The center of this province is in Nevada, the northern border is in southern Oregon and Idaho, the western border is the Sierra Nevada, and the eastern edge is the Wasatch Front in Utah. The western edge of Arizona and most of the southern one-third of the state are in this province (Map 2.3), and it extends eastward into New Mexico and southward into Mexico.

In the whole of this large Basin and Range Province there are well over 150 mountain ranges. Almost all of the ranges are long and narrow, they usually trend in a north-south direction, and they are separated by large, open basins. All these mountains were formed essentially in the same way, and they look alike, but their rocks vary tremendously and differ in age. Some areas are also highly mineralized. This is generally a dry region, but most of the erosion that weathered the mountains was caused by water, not wind. It is also an area with a great deal of internal drainage, where the water flows into basins that have no outlet to the sea (Figure 2.10). The best example of internal drainage is Great Salt Lake, but there are two rather good-sized basins of interior drainage in Arizona, Red Lake in the northwest and Wilcox Dry Lake in the southeast (Map 2.2). Such a dry lake is called a playa.

The Basin and Range Province in Arizona is characterized by broad, flat valleys (the basins) between narrow mountain ranges that seem to dominate the landscape. The mountains can be seen almost everywhere, but in the west they occupy only about 5 to 10 percent of the total area. In the southeast they are more prominent, occupying about 20 percent of the total

FIGURE 2.10. A typical scene in the Basin and Range Province of southern Arizona. It is open and flat, but mountains can be seen in the distance. In the right background is a playa (dry lake).

area. Most of the mountains in this region extend from south-southeast to north-northwest. They are not particularly high, only a few peaks are over 4,000 ft. (1,219 m.), but there are exceptions in southeastern Arizona where the mountains are of considerable height. The Santa Catalina Mountains, for example, reach an altitude of over 9,000 ft. (2,743 m.), and the Pinaleno Mountains are well over 10,000 ft. (3,048 m.). These mountains are all the more impressive because they begin from a low altitude, as opposed to the mountains that sit on top of the Colorado Plateau. Although most Basin and Range mountains are not high, they are quite rough, rugged, and difficult to cross, so highways have generally been built around them.

Most mountains of the Basin and Range Province were formed in the same manner. The rocks in this area were once relatively flat, but faulting caused them to break and to be pushed slowly upward along the fault lines (Figure 2.11). This action resulted in long, narrow fault-block mountains that have one side steeper than the other, and that steeper side is usually the fault side.

The basins between the mountains have been filled with alluvial material. The basins are not flat but gently rise from the center of the valley, where a dry wash is usually found, 1,000 to 2,000 ft. (305 to 610 m.) to the outer edges where the mountains begin. The basins are the lowest areas in Arizona in elevation, varying between approximately 100 ft. (30 m.) above sea level near Yuma to over 5,000 ft. (1,524 m.) in the southeastern part of the state.

At the fringes of the mountains, there are sloping areas where rocks are being eroded and transported toward the basin centers. Alluvial fans, which usually fringe a mountain to form what is termed a *bajada,* are found where small washes exit the mountains. These *bajadas* are usually desirable home sites in Arizona. Another feature common to the fringes of many mountain ranges are pediments (Figure 2.11), which are rock ledges that extend outward from the mountains. Pediments are usually covered with rock and dirt, but bedrock is found only a few feet below the surface.

The Basin and Range Province in Arizona can be divided into four sections. There are no really clear-cut boundaries to separate them, and they blend one into another. Three of the subregions—the Mojave Desert, the Colorado Desert, and the Gila Semi-Desert—are identified by vegetation and climate. This is not a particularly good way to identify regions, but there is little distinction in the topography to separate one region from the next. The fourth region, the Mexican Highlands, in southeastern Arizona, is physically quite different from the rest of the Basin and Range Province in Arizona.

The Mojave and Colorado deserts are basically the same, except for their vegetation. In crossing them, one gets the impression of open, flat deserts with jagged and serrated mountains always on the horizon. The Mojave Desert is long and narrow, extending from the Utah border to the Bill Williams River. The vegetation of this region is a spillover of that found in

SIMPLIFIED PROFILE OF STRUCTURAL FEATURES IN THE BASIN AND RANGE PROVINCE

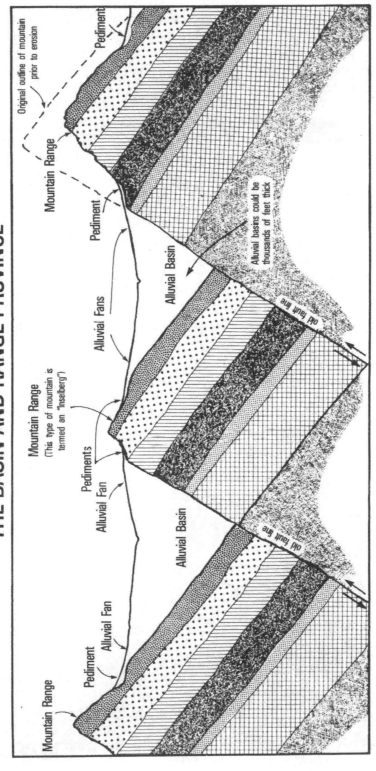

FIGURE 2.11

Nevada, and it is characterized by the Joshua tree. The Colorado Desert in southwestern Arizona looks like the Mojave Desert except for the different vegetation, which is a continuation of that of the Sonoran Desert to the south. These two areas are also sparsely populated, and most of the economic activity and towns are along the Colorado River. Exceptions to that rule are Ajo, a mining town, and Kingman, a county seat that is located at an important highway intersection. Many of the mountains of these areas are highly mineralized, and the areas have a history of mining activity.

The Gila Semi-Desert is east of the Colorado Desert, and Phoenix and Tucson are on its eastern edge. Geologically, this area resembles the first two areas, but it gets a bit more rainfall and is at a higher elevation, so the vegetation is more abundant. This is an area of large cities and many small towns, manufacturing, and government employment, and it has both surface water and groundwater for agriculture. This region dominates the rest of the state.

The area of the Mexican Highlands in southeastern Arizona is different from the rest of the Basin and Range Province in that the mountains are higher and occupy a larger percentage of the total area. This area also differs from other portions of the Basin and Range Province in that the average elevation is higher and it receives more rainfall. As a result, many people consider this area a part of the Mountain District of Arizona. It is, however, geologically very much a part of the Basin and Range Province, and it should be considered as such. This area has open vistas, forests along the streams and in the mountains, and rich grasslands and savannahs (Figure 2.12). It has always been an appealing landscape, and it is considered the best ranchland in the state. This region, especially the southern part, is also noted for its mineral production.

FIGURE 2.12. Open grassland in southeastern Arizona.

There are three large valleys in the Mexican Highlands, and they extend in a southeast to northwest direction and separate the area into districts. The first of these valleys, going east to west, is the San Simon, which extends into New Mexico. The second major valley is the Sulphur Springs Valley. Drainage in both of these valleys is northward, but most water in the Sulphur Springs Valley ends up in the Wilcox Dry Lake instead of going into the Gila River.

Between these two valleys are the Chiricahua Mountains, a major mountain mass. These mountains trend north-south, and a northern extension, the Dos Cabezas Mountains, swings to the west. Between these two mountain ranges is Apache Pass, which was important for early travelers. It was a dangerous pass, as the name implies, so the military established Fort Bowie to protect the travelers. It is an easy pass to cross, but there is a gentle gradient of several miles leading up to the pass, and when the railroad was extended into Arizona, it was more convenient to simply build it northward around the Dos Cabezas. Highways followed the lead of the railroad, and only a gravel road now goes through the once important Apache Pass. A northward extension of the Sulphur Springs Valley is the Aravaipa Valley, and between it and the San Simon Valley are the very high Pinaleno Mountains, with Mount Graham as the highest point, reaching over 10,700 ft. (3,261 m.). The third of the major valleys is that followed by the San Pedro River, which has a dependable flow along its upper reaches, and there is some agricultural development along it. West of the San Pedro are other high mountains, notably the Huachuca, Santa Rita, Rincon, and Santa Catalina mountains.

Central Mountain District

The Mountain District of Arizona is a transition zone between the Plateau and the Basin and Range regions. This is a very rugged area with many mountain groups. The mountains are not particularly high, the highest rises only to about the height of the Mogollon Rim, and they are not in tiers, as in the Basin and Range Province, but are a jumbled aggregate having little sense of symmetry to them. One mountain group often blends into the next, such as the Bradshaws into the Weaver Mountains, with no clear distinction where one ends and the next begins. The mountains in this region are of almost every type. Some were formed by faulting that began in late tertiary times, others were caused by folding, and yet others were formed by volcanic action. The rocks are also of every type, from sedimentary to volcanic to metamorphic.

There are several important mountain groups in the district, often separated by large valleys, and the major groups, from east to west, are the Sierra Ancha, the Mazatzal, and the Bradshaw mountains. Precipitation falling in the district flows southward to the Gila, Salt, and Bill Williams rivers. Local down-faulting and erosion by the streams has resulted in many valleys like the Tonto Basin and the upper Verde Valley. They are quite large and are known as the central valleys of Arizona.

The Bradshaw Mountains (Map 2.1) are typical of the central Mountain District. They are a large, irregular mountain mass, about 30 mi. (48 km.) wide and 45 mi. (72 km.) long. The highest point, roughly 8,000 ft. (2,438 m.), stands about a mile above the valleys at the mountain base. The rocks are mostly Precambrian schist into which granite masses have intruded. Other rock types are also found, and basaltic lava flows are superimposed upon much of the area. The lava has resisted erosion, so it stands out and often caps the mesas. The schist is the least resistant, so many valleys develop where this rock lies, and the more resistant granite rises as ridges and peaks.

This central Mountain District is important in several respects. It is a mineralized area, and mines have been in operation for a long time. It contains a few towns and some small cities, such as Prescott, Globe, and Payson, and the area is also very important as a watershed for the agriculture in the Salt River Valley.

CLIMATE

Climate is studied in terms of various climatic elements, such as temperature, wind, atmospheric pressure, and humidity (including clouds, precipitation, and evaporation). Arizona is very fortunate in having many climatic areas, varying from a subtropical desert on the lower Colorado to arctic conditions existing on top of the San Francisco Mountains. Many factors determine why the various climates are found where they are. Altitude is far and away the major factor influencing climate in Arizona, but altitude is not the sole determinant of the climate of a particular area. The higher elevations are usually cooler and wetter than the lower elevations, but there are many exceptions to that rule. For example, much of the Colorado Plateau is 5,000 ft. (1,524 m.) higher than the desert to the south, but summer on the Plateau can get almost as hot as on the southern desert, and the Plateau is certainly as dry as most of that southern desert.

Nevertheless, it is usually considerably cooler on the top of a mountain than at the base. If the air were still and one were to climb a mountain, it would usually be 3.3°F (1.8°C) cooler for every rise of 1,000 ft. (305 m.). This change in temperature is called the normal lapse rate. However, air is seldom still, and often it is moving upward to cross a mountain. In such a situation, the air changes temperature much more rapidly, and this is known as the adiabatic rate. As the air rises, it cools at 5.5°F (3.1°C) per 1,000 ft. (305 m.) until it reaches the dew point. This is known as the dry adiabatic rate. Once the air cools to the point at which condensation occurs within it, it cools more slowly (a varying rate, depending on local situations), and this is called the wet adiabatic rate. Moisture is therefore more effective on the top of a mountain than at the base; at the top it is cooler, the relative humidity is higher, and there is often a cloud cover. An annual rainfall of 20 in. (50.8 cm.) in such a situation results in a forest.

Air descending the other side of a mountain heats at the dry adiabatic rate, 5.5°F (3.1°C) per 1,000 ft. (305 m.). Air loses moisture crossing a mountain, but as it descends and warms it can again hold more and more. The relative humidity is, therefore, very low as it descends, and very dry desert conditions can prevail. This fact largely explains why much of the Plateau is dry even though it is high in elevation. It is surrounded by mountains, such as the Wasatch and the Mogollon Rim, and moisture is wrung from the air as it crosses them. An example of this climatic factor can be seen when traveling northward from Payson to Winslow. As one crosses the Rim, the land is forested, but as one descends to the Little Colorado River, the land becomes drier and drier until desert conditions are met, although the elevation is almost 5,000 ft. (1,524 m.).

Precipitation

Rainfall is always caused by a mass of air that cools, and the way an air mass cools is by rising. As rising air cools, the relative humidity changes because cool air cannot hold as much moisture as warm air. Soon the dew point is reached, clouds form, and precipitation falls (rain, snow, hail, etc.). The only other source of precipitation is the formation of dew and fog when an air mass right next to the surface cools, but those forms of precipitation contribute insignificant amounts of usable moisture in Arizona. To get rainfall, a large mass of air must rise rapidly. Of course, if the air mass is initially very humid, it will produce more rainfall, but even a relatively dry air mass, if it rises enough, will produce rain. There are three types of activity that will force air to rise; convectional, orographic, and cyclonic or frontal. All of the world's significant precipitation comes from one or a combination of these three conditions.

Convectional rainfall is the type of rainfall usually received in the summer months since it requires hot temperatures to make it begin. The earth on a summer day will receive much heat from the sun, and slowly the rocks and soil heat. This heat is transferred to the air just about at the surface, and soon that air gets very warm. No one knows what actually triggers it, but this increasingly warm air will burst through the cooler air above it and rise very quickly, at the rate of several thousand feet in a matter of a few minutes. As this air rises, it first cools at the dry adiabatic rate and at the slower wet adiabatic rate after the dew point is reached.

This type of activity produces violent thunderstorms (Figure 2.13). A large amount of rain can come from one of these thundershowers, easily 1–2 in. (2.5–5.1 cm.), as well as a great deal of hail and thunder and lightning (Figure 2.14). It will not be a widespread rain, and an area immediately next to the track of a storm will receive nothing but wind. These thunderstorms

FIGURE 2.13. Thunderhead over the San Carlos Indian Reservation.

can also cause much damage, because so much rain falls in such a short time that the soil cannot absorb the water and it runs off, causing erosion and a flooding of the washes. These storms usually begin about two or three o'clock in the afternoon, after the air has had sufficient time to heat. Once a thunderstorm forms, it will usually last until about midnight, at which time the air being drawn into it will be relatively cool.

A lot of dust is often associated with these thunderstorms. There will be a great deal of dust in the air around thunderheads, since the winds blowing into the storm will pick up dust. Another type of dust associated with the thunderstorm is a haboob, and Phoenix will average several such dust storms each year. Air that rises in a thunderhead eventually gets very cold, and often it descends en masse toward the earth many miles from the thunderhead. The air heats as its descends, but it is still relatively cool compared to the surrounding air, and a drop of 23°F (12.8°C) has been reported after such a storm passes. This mass of air will kick up a great deal of dust and produce a haboob (Figure 2.15). The air mass can carry dust quite high, up to 10,000 ft. (3,047 m.), and it will advance like a front, even over small mountains. Haboobs will advance at up to 45 miles per hour (72.4 km./hr.), and the highest winds within one will vary between 60 and 90 miles per hour (96.5 and 145 km./hr.). These extensive dust storms are seldom serious and cause few problems because the desert surface, through time, develops a hard crust and little loose dirt can be picked up. However, when that hard

FIGURE 2.14 . Rain falling from the bottom of a thunderstorm.

desert crust is broken by the action of man, for example, by dirt bikes or dune buggies, a high wind will pick up a large amount of dust and reduce visibility to near zero. Serious highway accidents can occur as a result.

Orographic rainfall occurs when the winds are forced to rise to cross a mountain barrier, which cools them rapidly. The type and amount of precipitation that results depends on the height of the mountain, the moisture content of the air, and the temperature of the air mass. The windward side of the mountain will receive more moisture than the leeward side, which is said to be in "the rain shadow." Orographic rainfall can occur throughout the year in Arizona, and it occurs whenever winds with any moisture are forced to cross the central mountains.

Cyclonic rainfall is the result of the movement of large air masses, and it is quite complex. This condition produces most winter rainfall. Anyone who follows the weather reports realizes that a series of highs and lows sweeps across the United States from west to east. The highs are known as anti-cyclones and produce clear weather. The lows or cyclones have just the op-posite effect; they are areas of low pressure, with rising unstable air in the center and winds moving inward. Cyclonic storms usually follow a storm belt, or storm track, across the United States from the Pacific Northwest toward the Midwest. Most of these storms pass north of Arizona, but every winter several enter the state and bring moisture. The location of the storm track appears to be determined by the jet stream, and about once in every 20

32

SIMPLIFIED CROSS SECTION OF THUNDERHEAD AND HABOOB

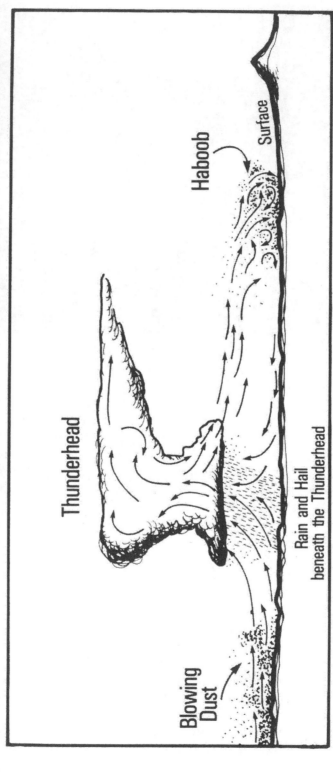

FIGURE 2.15

years the jet stream will shift southward for most of the winter. In such a winter, Arizona will get considerably more rain than the normally wet state of Washington. Another cyclonic condition that should be mentioned is one that occasionally occurs in the fall months. About once every 15 or 20 years, a large hurricane will develop off the coast of Baja California and move northward into Arizona. By the time it reaches Arizona, it is no longer classified as a hurricane but a "tropical depression." It will, however, bring the state heavy amounts of rain.

Another type of winter rainfall closely associated with the movement of cyclonic storms and large air masses is frontal rainfall. Frontal precipitation occurs when two differing air masses are next to one another and the warmer air mass is forced to rise over the cooler one. The rising warmer air cools, which produces rain, sleet, or snow depending on local conditions.

Winter precipitation, produced by cyclonic or frontal conditions, is very different from summer rains. It often takes the form of a slow drizzle that lasts for days, and it can result in a heavy snowfall in the mountains (Figure 2.16). This is the best type of precipitation, for the moisture will slowly be absorbed by the soil and there will be a good runoff. After a series of good winter rains, cattlemen, lumbermen, and farmers look forward to a good year.

It must be remembered that rainfall in the state often comes from a combination of factors. A large cyclonic condition may enter the state but cause rain only in the mountains when forced to cross that district. Another exam-

FIGURE 2.16. A typical winter storm near Sedona; there are dark, widespread clouds that are low to the earth.

ple is when heated air from the desert moves upslope into the Mountain District, resulting in almost daily thundershowers in that area.

Arizona has two wetter parts of the year, summer and winter, and two drier, spring and fall (Figure 2.17). The summer rains occur mostly in July and August; those months have been termed the monsoon season by meteorologists, and they are usually the wettest months in all parts of the state. June is often hot enough to trigger thunderstorms, but the humidity is too low. By July, however, the relative humidity increases as humid air moves northward from the Gulf of California, and the thunderstorm season begins. December to March are the months of the winter storms. About half of the moisture the state receives occurs during this season, and it is the type of slow rain most desired, because it soaks in and causes the least amount of erosion. There are exceptions to these broad generalizations: for example, the western part of the state especially benefits from winter rains, but the northeast is exceptionally dry in winter because it is surrounded by moisture-trapping mountains.

The driest months are in the late spring, particularly in the months of May

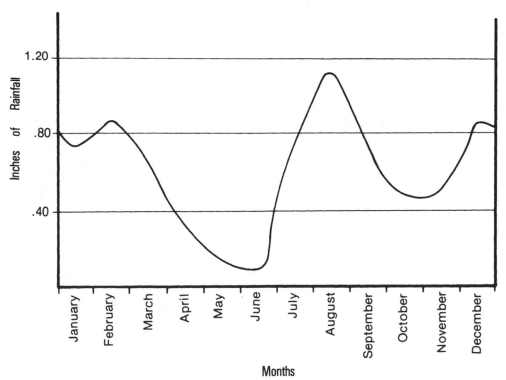

FIGURE 2.17

and June. It is a good time of the year to plan outdoor activities, because it is too late in the year for the winter storms and too early for the thunderstorms. There is another dry period in the fall, centered on the months of October and November, but that season is not nearly as dry as the spring months, and it is the time of year when an occasional tropical depression will enter the state.

The average distribution of precipitation in the state is shown on Map 2.4, and several major factors can easily be seen. Most of the mountain regions of the state stand out as receiving up to or more than 25 in. (63.5 cm) of rainfall a year, such as the Bradshaws, the White Mountains, and most of the mountains in the southeast. Much of the Plateau, however, although at a

AVERAGE ANNUAL PRECIPITATION

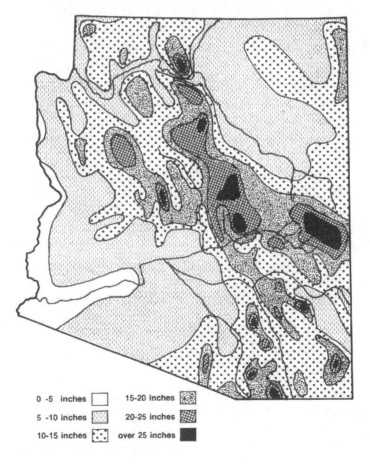

0 -5 inches   15-20 inches

5 -10 inches   20-25 inches

10-15 inches   over 25 inches

MAP 2.4
Source: C. R. Green and W. Sellers, eds., *Arizona Climate*
(Tucson: University of Arizona Press, 1964).

high elevation, receives less than 10 in. (25.4 cm.) of rainfall a year. It must be remembered that these figures are only averages and the total amount of rainfall a place may receive will vary greatly from one year to the next.

Temperature

Arizona is known for its temperature extremes. In the deserts, particularly to the south and west, the summer days get very hot. This region has a low elevation, and there is no cloud cover. Georgia lies at almost the same latitude as Arizona, but Georgia has a higher relative humidity and more cloud cover. The clouds reflect sunlight back into space, whereas in Arizona the sun heats up the exposed rock and soil, and they transmit their heat to the air just above them. In the summer, newcomers think the southern and western deserts resemble an oven. It gets particularly hot along the Colorado River, and all the towns in that area have reached 120°F (48.9°C); the record is 127°F (52.8°C), and two towns have hit that. However, the nights do cool considerably, and a difference of 40°F (22°C) between day and night temperatures is not unusual. This temperature extreme, from day to night, is known as the diurnal temperature range. The drop in temperature is so great because the deserts radiate off much of their heat and there is no cloud cover to act as a blanket to keep that heat near the earth. The relative humidity must be low for these great diurnal temperature ranges to occur.

The southern deserts get very hot in the summer, and the mountain sections of the state get bitterly cold in the winter. It is not unusual in the wintertime for Arizona to have the hottest and coldest spots in the contiguous 48 states on the same day. The lowest temperature ever recorded in the state was at Hawley Lake in the White Mountains, where it once reached −41°F (−40.5°C).

The climate of the urban areas of the state is changing, and not for the better. In those areas, there is a heat island effect because energy from the sun is absorbed by the buildings and other concrete and asphalt surfaces and then is released at night. There is also a greenhouse effect because all the dust and pollutants in the air prevent the heat from escaping at night. Thus evenings are not cooling as much as they once did, and the average night-time temperatures of the cities, expecially Phoenix, is rising. The greenhouse effect may also keep the days cooler in the wintertime, by preventing some of the sun's heat from reaching the surface. The summer daytime temperatures in the cities are also rising; and another aspect of this higher temperature is more and stronger winds in the cities. There is much research yet to be done on how cities affect their local climates, but none of the effects are beneficial.

Associated with high temperatures and low humidity is increased evaporation. The evaporation and evapotranspiration rates in Arizona are highest along the western border. That area receives the least cloud cover, which means it gets the greatest amount of solar radiation, and that fact, coupled with the area's low average elevation, results in the highest temperatures in the state. A great deal of moisture is thus lost to the atmosphere by evaporation (Map 2.5). If a pan of water were placed outside, and water added as the water in the pan evaporated, the potential for evaporation would be found to be much greater than the average amount of total rainfall. In Phoenix, for example, 110 in. (279.4 cm.) would evaporate in a year, and in the average year only 7 in. (17.8 cm.) of rain would fall. Added to the evaporation is the moisture given off by plants; in the Phoenix area, plants have the potential of transpiring an additional 40 in. (101.6 cm.) of water a year (Map 2.6). A combination of the two rates shows why Arizona is mostly desert.

AVERAGE ANNUAL PAN EVAPORATION

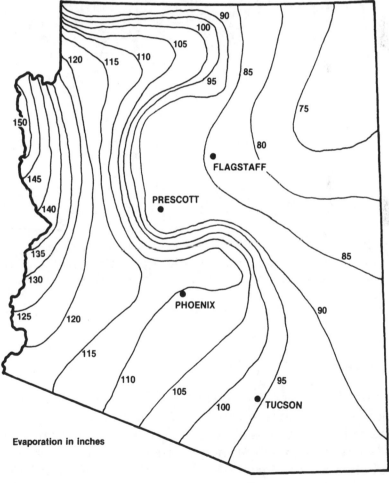

MAP 2.5
Source: Laboratory of Climatology, Arizona State University, "Arizona Resources Information System," Cooperative Publication No. 5, 1975.

AVERAGE ANNUAL POTENTIAL EVAPOTRANSPIRATION

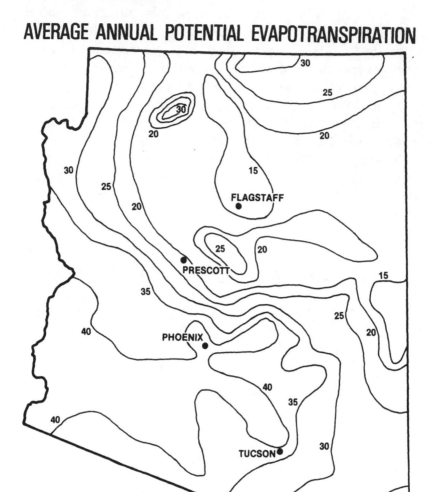

MAP 2.6
Source: Laboratory of Climatology, Arizona State University, "Arizona Resources Information System," Cooperative Publication No. 5, 1975.

VEGETATION

Many factors determine the vegetation in Arizona. Some such as precipitation, altitude, and temperature are quite obvious, but on a smaller scale, factors such as soil type, slope, and exposure are just as important. For example, there is little recognition of the importance of exposure as a determinant, but often in Arizona the north side of a hill will have different vegetation than the south. A south-facing slope is drier and warmer, because it receives the more direct rays of the sun, and a north-facing slope will have much lusher vegetation, because it is damper and cooler. The tree line on a slope with a northern exposure is about 11,000 ft. (3,353 m.), but on a much warmer and drier south-facing slope the tree line is considerably higher, about 11,500 ft. (3,505 m.).

Also important in understanding vegetation and vegetation change in Arizona are the actions of humans, but just how much impact humans have had is a very controversial topic. Most authorities believe the early Indians had little effect on the vegetation. They undoubtedly used fire as a hunting aid, and they did clear land for agriculture, but otherwise their impact was probably minimal.

But it is next to impossible to use the written accounts and descriptions of the early visitors in order to discern what changes have taken place since the time of the early Indians. Some of the first Spaniards to visit the area described it in very favorable terms, and some parts that they described as having high grass are now obviously rather barren desert. But can the early accounts be trusted? Some writers may have wanted to please the higher authority to which they wrote and therefore may have lied, or at least stretched the truth greatly, or perhaps they were describing an area in a wet spring after a series of wet years. Also, one writer may describe an area as being rich in grass, and another may describe the same area as being a barren desert.

Several factors may account for the variety of opinions, but the most important factor to consider is the culture from which the writer came. The area that is now southern Arizona, especially the southeastern part, probably was such a pleasing sight to early Spaniards that some may have thought they had died and gone to heaven. In the summertime the Spanish Meseta, which dominates central Spain and which would have been familiar to many of the Spaniards, greatly resembles the Sahara Desert because of its intense heat and dry conditions. To a Spaniard then, the area that is now southern Arizona—with its beautiful hills of rolling grass, mountains edged with forests and many springs, and small permanent streams in the valley bottoms—would have seemed an excellent environment. There was plenty of water in small oases and more than enough grass for animals to graze. It is little wonder that most Spanish accounts were written in such positive terms.

Americans coming from the East, however, had a different opinion of that area. Most considered it barren, devoid of good soil, unbearably hot, and unfit for human habitation, and many eastern writers said so in no uncertain terms. Those writers came from a humid region, and they could not understand or appreciate such an "alien" environment. One example was Gen. William Tecumseh Sherman who called the area "an immense, miserable country" and a "God-forsaken land"—although he had never visited it! (U.S., Congress, House, Committee on Military Affairs, *Report to Accompany the Bill (H.R. 2546) to Provide for the Gradual Reduction of the Army of the United States*, H. Rept. 384, 43rd Cong., 1st sess. [Washington, D.C.: Government

Printing Office, 1874], pp. 6, 19. In the same report, Sherman made the facetious comment, "If you, gentlemen, will get Mexico to take Arizona back I will agree to knock two regiments of cavalry from our estimates" [p. 5].)

There have been some rather obvious vegetation changes in recent years, and two changes that are often cited are the spread of juniper and piñon pine in the north and the spread of mesquite in the south. All are small, bushy trees of limited value, and they have been invading the grasslands. Perhaps the best way to see how much the vegetation has changed is to compare early photographs with recent ones, even though photography in Arizona goes back to only the late 1800s. Such a comparison of photographs of southern Arizona shows there have been changes, especially a decline of the grasslands and a spreading of shrubs.

The exact causes for these changes are really unknown, but any number of reasons could be put forward to explain the spread of these trees, such as overgrazing and the lack of frequent burning (which was probably very common prior to the arrival of the Europeans). Another commonly accepted view is that grazing animals (cattle and sheep) replaced browsers (deer), and the result was a loss of grass and a lack of fires that burned the grass and thus kept the shrubs in check. A common argument given for the spread of mesquite is that cattle eat the mesquite beans, the beans pass through the animals' alimentary tracts unharmed, and the beans are then deposited in fresh fertilizer ready to germinate and grow. A small variation in climate may explain the change; it has also been proposed that change is natural in that there is no real "climax" vegetation and plant dominance may come and go in cycles. Man may also be the responsible agent because southernmost Arizona has been used extensively for grazing for about 200 years by Spaniards, Mexicans, and Americans. Perhaps overgrazing has destroyed the grass cover, encouraging erosion and the spread of the mesquite; the mesquite, in turn, has encouraged more erosion and the exposure of more and more bare rock; this has changed the albedo rate, which has slowly changed the climate. Many argue that this process of desertification has occurred in the Sahara and that it is still taking place there and beginning in Arizona. That may be true, but the desertification may also be part of a natural process that began in the late ice age when the climate of the Sahara began to change. There is, however, no single answer to explain why some plants are expanding their range; the expansion is the result of an interplay of many factors, many of which are not understood and others are probably unknown. Obviously, much more study needs to be done along these lines.

There is tremendous variation in the vegetation of Arizona, ranging from barren deserts to arctic conditions. The many types of vegetation in Arizona can be divided into five main groups: alpine, forest, grassland, desert, and imported or "exotic."

Alpine Vegetation

Alpine vegetation in Arizona is found only in the San Francisco Mountains, beginning at about 11,500 ft. (3,505 m.) and continuing to the top. Baldy Peak in the White Mountains is a little over 11,500 ft. (3,505 m.), and vegetation there is very close to being alpine, but not quite. Alpine vegetation exists only above the tree line, and the type of vegetation found there resembles that located above the Arctic Circle. Vegetation in this zone is mostly sedges, mosses, forbs, grasses, and lichens.

Forest Vegetation

Forest vegetation is found where the soil is moist throughout most of the year, especially during the growing season. If the soil goes through a regular dry period during the year, it will probably support grass or possibly a chaparral forest. Some soil can be too wet for forests, such as along the streams in northern Arizona, and when this is the case, meadows are found (Figure 2.18).

The forest vegetation of Arizona can be subdivided into several groups, each based on the dominant vegetation: subalpine, ponderosa pine, piñon-juniper, chaparral, oak, and gallery (Map 2.7). A subalpine forest grows just below the tree line (Figure 2.19), and such forests are found around the San Francisco Mountains, in the higher parts of the White Mountains, and on the Kaibab Plateau. The major tree types are Engelmann spruce and two types of firs, alpine and corkbark. Now and then aspens are seen, and Douglas firs are found along the lower limits of a subalpine forest.

The ponderosa pine forest is the most extensive type of forest in the state, and it is far and away the most valuable economically, since almost all of the trees cut for lumber in the state are ponderosa pine. Pure stands of ponderosa occur below the subalpine forest of spruce and fir, and a ponderosa pine forest is commonly found at elevations of approximately 6,000 ft. to 8,500 ft. (1,829 m. to 2,591 m.). The largest concentration of ponderosa in the world is found along the Mogollon Rim, extending in an unbroken stand from the New Mexico border almost to Williams, Arizona. Outliers of this forest are found in other areas, especially the higher peaks in the northeast, the Kaibab Plateau, the mountains in the central portion of the state, and even some mountaintops in southern Arizona, such as the Chiricahua and the Santa Catalina mountains. The pine forest is often quite open, and frequently there is a ground cover of grass used for grazing (Figure 2.20). In other areas, there is a dense understory of shrubby plants instead of grass, and their role in forest ecology is not yet understood.

It is also not known what role fire plays in this forest ecology — or in the types of forests described below. Fires can be extremely destructive, but

FIGURE 2.18. A meadow in the White Mountains.

FORESTS

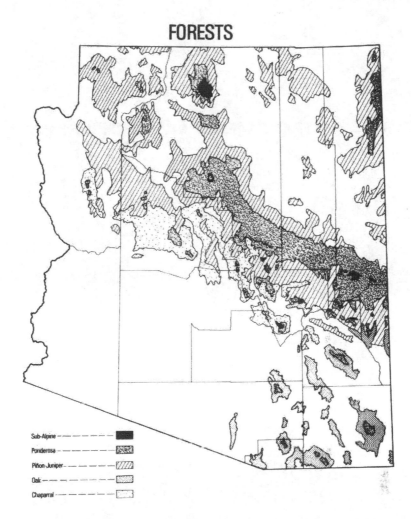

Sub-Alpine
Ponderosa
Piñon-Juniper
Oak
Chaparral

MAP 2.7
Source: Arizona Resources Information System, "The Natural Vegetative Communities of Arizona," 1973.

Americans seem to have a "Smokey the Bear" attitude that all fires are harmful. This is not true, and it is common for a fire to do more good than harm if it is allowed to burn. These fires are small ground fires that consume the grass and bushes, and perhaps, if allowed to burn, they would eliminate the small shrubs found in some of the ponderosa forests. Burning this underbrush before it gets too dense also helps prevent truly large forest fires later and helps maintain a forest dominated by pine trees.

The piñon-juniper forests are found at lower elevations than the ponderosa pine forests, anywhere from 4,500 to 7,500 ft. (1,372 m. to 2,286 m.), and in areas that receive less moisture. It is easy to see this change as one leaves the ponderosa forest: the ponderosa trees begin to thin out and

FIGURE 2.19. A typical subalpine forest in the White Mountains.

FIGURE 2.20. Typical ponderosa forest scene. Note the wide spacing of the trees and the extensive grass cover.

become shorter, other smaller types of trees, piñon and juniper, begin to replace them, and soon one is in a different type of forest. Piñon and juniper are both small, bushy trees that are of a very limited value economically (Figure 2.21). This forest type is a transition to grassland, but often there is no sharp boundary between the two; these small trees will often be found in a grassland, and where the trees are widely dispersed, large areas of grassland can be found in this type of forest. These trees are now considered little better than weeds, and their elimination is a goal of range management. They use much precious water that otherwise could support grass, and grass cover would result in more runoff, less erosion, and more grazing land for cattle. There are several ways to kill these types of trees, but the most conspicuous is to have two very large tractors drag a heavy chain across the land.

Chaparral is found in many scattered locations south and west of the Mogollon Rim. The single largest stand of this type of vegetation is found south of the westernmost edge of the Mogollon Rim, but patches of chaparral are found in other areas, such as all along the highway between Superior and Globe. This vegetation type is also a transition to grassland, and these plants are somewhat drought resistant. This vegetation type is very different from the piñon-juniper forest, which is coniferous, because it is made up of various broad-leafed sclerophyllous shrubs. The vegetation consists of many types of small trees and shrubs, such as scrub oak, manzanita, buckthorn, jojoba, sumac, Apache plume, yerba santa, mountain mahogany, and many others. The chaparral in Arizona is very different from that found in California, and they really should not be compared. The chaparral in Arizona is of little economic use, and range management officials would like it eliminated and replaced with grass. A grass cover would result in less erosion, less silting of lakes and streams, and greater runoff. Powerful herbicides, such as

FIGURE 2.21. A piñon-juniper forest. This was formerly a grassland into which juniper especially has been expanding.

Kuron, are sometimes sprayed from the air in an attempt to kill the chaparral, but such actions are dangerous to humans and livestock. California has been very successful in the use of goats to maintain firebreaks in the chaparral areas, and a similar large-scale use of goats is one technique that may help control the chaparral in Arizona.

Oak forests are found in scattered spots in southeast Arizona, and they, too, are a transition between forest and grassland. Oak forests are particularly prominent at the higher elevations in the Mexican Highlands, just below the ponderosa pine forests. The greatest development of oak forests is in the Sierra Madre Occidental in northwestern Mexico, and one area of their greatest development in the United States is in Arizona near the Mexican border. These oak forests take on the look of savannahs or open park landscapes, with much grass and many scattered oak trees. Today, mesquite is invading many of the oak forests, and the oaks are retreating.

The last type of forest in Arizona is the gallery forest (Figure 2.22). This is a forest along a stream course in a desert or grassland area, and gallery forests resemble thin fingers of forest growth extending across an otherwise open landscape. Trees in this type of forest are often water-loving trees, such as Arizona sycamore or cottonwood, but in the drier washes there are also other types, such as mesquite and desert willow. This type of forest is

FIGURE 2.22. A gallery forest of cottonwood and sycamore along Arivaipa Creek. Large cottonwoods and sycamores follow the creek bed, but away from the stream only desert vegetation is found.

considered a very important habitat for wildlife and therefore is significant in the overall ecology.

Grassland Vegetation

Grassland areas in Arizona are usually at lower elevations and in drier environments than the forests. These grasses produce excellent forage, because they are very nutritious. Looking out over the grasslands of Arizona, the whole area seems to be thickly covered with grass, but if one looks down, it is apparent that the grass grows in clumps and some of the surface is bare.

The grasslands of Arizona are divided into two types (Map 2.8). The short-

GRASSLANDS

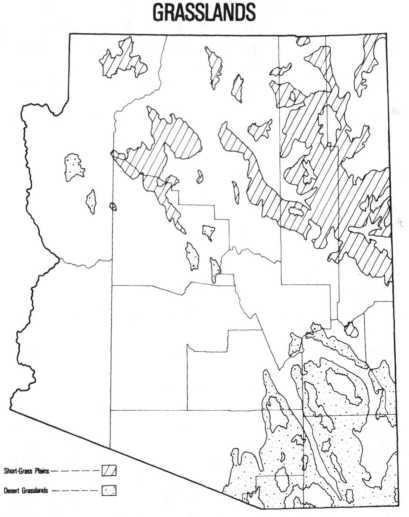

Short-Grass Plains — — — — — ▨

Desert Grasslands — — — — — ⬚

MAP 2.8
Source: Arizona Resources Information System, "The Natural Vegetative Communities of Arizona," 1973.

grass plains are grasslands found in the northern part of the state, and they lie between the piñon-juniper and desert zones (Figure 2.23). These grasslands are considered to be a western extension from the High Plains of the central United States, and there are some low passes in New Mexico where almost continuous stands of this grass can be found extending into Arizona. These grasses have adapted to the cold, dry, and harsh conditions, and they usually form pure stands so that travelers through the area see open and wide expanses of rolling short-grass plains. Grama grass dominates, especially blue grama, but other grama species, such as side oats and black grama, are also found. These short grasses sometimes occur as a thick carpet, but more usually they are found in open stands.

The desert grasslands are mostly found in the southern parts of the state, especially the southeast (Figure 2.24). They are typically located in broad belts around the bases of the mountain ranges. This type vegetation is considered to be an extension of the grasslands in Mexico, and it provides excellent forage for cattle. The most important grass type on this range is the grama grasses, and second in significance are various species within the genus *Hilaria* (particularly tobosa and curly mesquite), but other common grass genera include blue stems, three-awns, sprangletops, and many others. Various palatable shrubs are also found growing in this type of grassland, but frequently various woody shrubs, such as mesquite, cholla, palmilla, and other cacti, are mixed in with it. The desert grassland is the most arid of the North American grassland regions. Rainfall is between 10 and 15 in. (25.4 and 38.1 cm.), but usually a bit more than half of it falls in July, August, and early September. It is along the lower and drier edges of this environment that mesquite infestations have become a real problem (Figure 2.25).

Desert Vegetation

Arizona has two major types of desert. The one on the Plateau is called the Northern Desert (Map 2.9), and it is a southern extension of the Great Basin Desert centered in Utah and Nevada. Small scattered shrubs, such as sagebrush, blackbrush, mormon tea, and shadescale (saltbrush), are characteristic in this area. These shrubs are found at lower elevations on the Plateau, such as along the Little Colorado and Colorado rivers and along Kanab Creek. Few people live in this type of desert.

The desert of southern Arizona is the Sonoran Desert, and about one-third of Arizona is covered by this extension of the northwestern Mexican desert into the United States. In Arizona, this desert is divided into three districts—the Mojave Desert, the Colorado Desert, and the Gila Semi-Desert. This Sonoran Desert is an important area; the majority of Arizonans live in this environment, because the cities of Phoenix and Tucson are found within its bounds.

FIGURE 2.23. A scene in the northern grasslands. The dark patches on the mountains in the background are piñon-juniper forests.

FIGURE 2.24. Desert grassland. Note the large number of bushes and shrubs. This picture was taken near the center of the large Babocomari Grant in southeastern Arizona.

FIGURE 2.25. Large numbers of mesquite trees expanding into once open grassland.

Xerophytic plants capable of withstanding long drought are the rule in this hot, dry desert. Plants in this environment fight the drought in a wide variety of ways; some are microphyllous, which refers to their small and hard leaves, and some are succulents, which refers to their ability to store water. Some of the other techniques used by plants to survive in the desert include having a widespread root system, having a deep tap root, or having a waxy substance on the leaves to retard transpiration. The Sonoran Desert is noted for its abundant vegetation, and it has more vegetation than any other desert in the world. Many visitors come to Arizona with the idea that all deserts are like the sandy wastes of the Sahara, and they initially exclaim that Arizona is not a desert because of its vegetation. However, if those visitors were to spend a summer in that part of Arizona, they would soon agree that it is indeed a desert.

The vegetation in this desert is closely related to the topography. The rough slopes along the mountains are covered with cholla, prickly pear, ocotillo, paloverde, yucca, and other desert plants. Further down the slopes there are great concentrations of cacti, such as the tall saguaro, which are located mostly on the drier south-facing slopes. Out in the middle of the basins shrubs predominate, the creosote bush being particularly common.

DESERTS

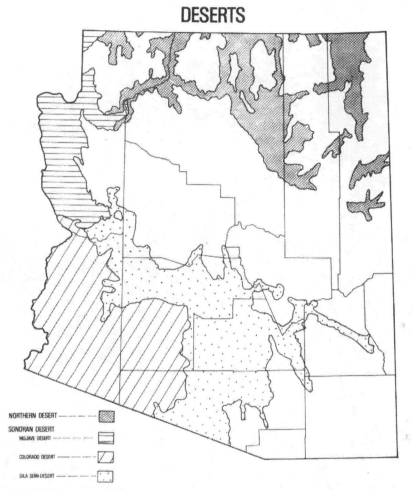

NORTHERN DESERT

SONORAN DESERT
   MOJAVE DESERT
   COLORADO DESERT
   GILA SEMI-DESERT

MAP 2.9
Source: Arizona Resources Information System, "The Natural Vegetative
Communities of Arizona," 1973.

Since basins occupy most of the area, thousands of square miles are covered by this rather ubiquitous plant. Many dry washes cross the basins,
and vegetation in the washes consists of trees such as mesquite and
desert willow. If a wash carries enough moisture, there will be a gallery
forest.

The vegetation of the three subregions of the Sonoran Desert of Arizona
varies greatly. The Gila Semi-Desert, located in south-central Arizona near
the major cities of Phoenix and Tucson, is quite rich in vegetation (Figure
2.26). The Colorado Desert in the extreme southwest is exceptionally dry.

FIGURE 2.26. Abundant vegetation in the Gila Semi-Desert as seen in the Super-stition Mountains east of Phoenix. This photo was taken in early spring.

This area receives about 5 in. (12.7 cm.) of rain a year, or less, and the vegetation is sparse (Figure 2.27). When traveling to Yuma from Phoenix or Tucson, one can easily see a change in the vegetation near Gila Bend, and from that point on the natural vegetation seems scanty and widespread. The Mojave Desert located in the westernmost part of the state, from about Parker northward to the Utah border, is a spillover of the Mojave Desert in California and Nevada. The Joshua tree is characteristic of this desert.

Exotic Vegetation

Exotic vegetation has been imported into the state by humans and is not native to Arizona. Instead of adjusting to and living with the local environment, many Americans who moved to Arizona, tried to reestablish, as best they could, the environment they had left. As a result, the cities in Arizona largely resemble eastern cities with their green lawns and trees. It almost seems that by planting greenery, Americans were asserting their dominance over a very harsh environment. A recent tendency toward "desert landscaping" is an environmentally sensitive reversal of the earlier trend.

Many of the exotic plants are beneficial, but some—for example, the Russian thistle (tumbleweed) and salt cedar—have turned out to be real pests. Grasses planted in the urban areas are almost invariably exotic, and the trees that have been introduced from almost every corner of the earth are very conspicuous. Some of the exotic trees are the tamarisk, from Asia; the ubiquitous mulberry, from China; the African sumac; and many varieties of eucalyptus, from Australia. The area of the world that has contributed the most, however, is the region around the Mediterranean. From this part of the world have come the aleppo pine, the corkbark oak, the carob, date palm, citrus (originally from Southeast Asia), and olive trees, and many others.

FIGURE 2.27. The Colorado Desert near Yuma. The vegetation is very sparse and mostly consists of creosote bushes with various other trees and bushes in the washes.

A North-South Profile of Vegetation

Since Arizona has such a great range of vegetation, one of the best ways to illustrate the zonation of the vegetation is with a profile of the state (Figure 2.28). The profile is greatly exaggerated, but it depicts what vegetation a person would see if one were to travel from the Mexican border in the area of Ajo, northward toward Flagstaff, and over the San Francisco Mountains to the Utah border near Page. Exact figures for when one vegetation type stops or starts are impossible to give, because one zone slowly grades into another and many local variables will affect vegetation (Figure 2.29).

55

A NORTH-SOUTH PROFILE OF ARIZONA'S VEGETATION ZONES

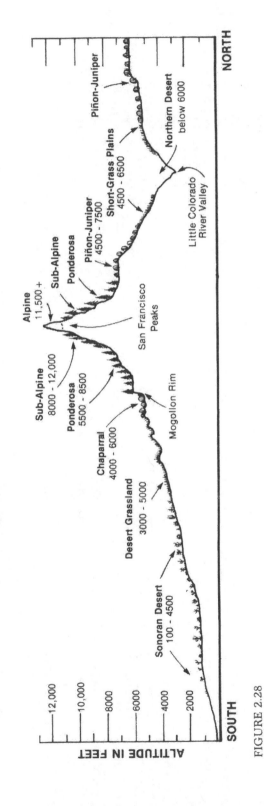

FIGURE 2.28

FIGURE 2.29

VEGETATION IN ARIZONA ACCORDING TO ELEVATION

Name of Vegetation Type	Characteristic Plants	Elevation (feet)	Rainfall (inches)	Percentages of Total Vegetation	Location in Arizona
Alpine tundra (Arctic-Alpine Life Zone)	Mountain avens, alpine sedges, and grasses	11,500–12,633	30–35		Above timberline on summit of San Francisco Mountains
		FORESTS			
Spruce-fir forest (sub-alpine forest, Hudsonian and Canadian Life Zones)	Engelmann spruce, alpine fir, corkbark fir	8,500–12,000	30–35	1	High elevations, White Mountains, San Francisco Mountains, and Kaibab Plateau
Douglas-fir forest (montane forest, Canadian Life Zone)	Douglas fir, white fir, quaking aspens, limber pine	8,000–9,500	25–30	2	High mountains in eastern and northern parts
Ponderosa pine forest (Transition Life Zone)	Ponderosa pine, Arizona pine	5,500–8,500	19–25	6	Mountains and plateaus in north-eastern half
Piñon-juniper woodland (Upper Sonoran Life Zone)	Piñon pines and various junipers	4,500–7,500	12–20	17	Plateaus and mountains in northern half
Chaparral-Oak woodland (Upper Sonoran Life Zone)	Shrub live oak, manzanitas, sumacs, cliff roses, ceanothuses	4,000–6,000	13–25	8	Mountains in central part

Type	Dominant plants	Elevation (ft)	Precipitation (in)	Percent	Location
GRASSLANDS					
Short-grass (plains grassland, Upper Sonoran Life Zone)	Blue grama, hairy grama, galleta, buffalo grass	4,500-6,500	9-20	15	Plains and plateaus in northern part
Desert grass (semidesert grassland, Lower Sonoran Life Zone)	Black grama, tobosa, dropseeds	3,000-5,000	9-18	10	Plains in southeastern part
DESERTS					
Sagebrush (northern desert, Upper Sonoran Life Zone)	Big sagebrush, black brush	2,500-6,000	7-17	6	Lower portions of northern area; Little Colorado River drainage
Desert (semidesert shrub, Lower Sonoran Life Zone)	Creosote bush, mesquite, tarbush, acacias, palc-verdes, bur sages, cacti, saltbush	100-4,500	3-15	35	Southwestern half of the state and bottom of Grand Canyon

Source: By permission from Arizona -- Its People and Resources, Faculty of the University of Arizona (Tucson: University of Arizona Press, 1972).

3

EARLY SETTLEMENT OF THE LAND

ARRIVAL OF INDIANS IN THE NEW WORLD

Indians have lived in the area that is now Arizona for a very long time, at least 15,000 years, whereas whites have been there only a very short time. The early Indians left very little trace of their civilization, and because they had only a slight impact upon the environment they are not a part of the heritage of most Arizonans. Yet the Indians did occupy the area for a long time and, therefore, should be studied in some detail.

The first Europeans who settled along the eastern seashore of North America were ignorant of basic geographic facts—such as the closeness of Asia to North America in the extreme Northwest—and of some of the basic concepts of anthropology. They did not know who the Indians were or from where they came. Later, when the Europeans crossed the Appalachian Mountains and entered the valleys of the Mississippi and Ohio rivers, they found large earthen structures known as Indian mounds. These mounds were very large, and the Cahokia Mounds outside of St. Louis are still the world's largest solid earth structures. The whites thought that surely the local Indians, or their ancestors, could never have built such structures, and they began searching for ideas that might explain the mounds.

Many ideas were put forward to explain who built the ancient mounds, including giants (or the "mysterious mound builders"), wandering Welshmen, German ironmongers, Irish monks, Vikings, and Phoenicians. The favorite explanation, and the one advocated by Pres. William Henry Harrison, was that the mounds were built by Hebrews, one of the ten lost tribes of Israel. These ideas may seem farfetched today, but at the time they sounded like logical answers to unresolvable questions. They were a product of the ignorance of the period and the strong religious ideas commonly held. There is, however, no evidence that North America was ever inhabited by anyone other than the ancestors of the modern American Indians.

The first people to come to the New World came by way of the Bering Strait. The last of the glacial stages, the Wisconsin, began about 70,000 years ago and lasted until about 10,000 years ago. As the ice built up, the seas were lowered at least 300 ft. (91.4 m.) and maybe as much as 450 ft. (137 m.) at the height of that ice age. When the seas were lowered, a land bridge up to 1,300 mi. (2,092 km.) wide formed between Asia and North America, and the two continents were no longer separated. There were two unglaciated corridors allowing easy movement between the two continents, one through central Alaska and the other along the Alaskan coast. Many kinds of animals moved across this land bridge, going both ways. Man also crossed at this time, and over a long period probably many different groups migrated to the New World.

No one knows the exact date when humans first came to the New World. There are many radiocarbon datings that tend to verify that humans were in the New World by 12,000 B.C. One site, at Naco near Bisbee, indicates that man was in the Arizona area as early as 9000 B.C. and perhaps considerably earlier. Although there are no earlier verifications, most authorities agree that 20,000 B.C. is a nice round figure to use when estimating when the first humans arrived in the New World.

By the time of the last glacial stage, humans had learned to live in a cold, harsh environment. They did not evolve to fit the Arctic, like the polar bear, but through their culture (the use of fire, shelter, clothing, etc.) they were able to live near the glaciers. This way of life, known as the Upper Paleolithic culture, was found in areas as widely scattered as North Africa, western Europe (humans had retreated from Europe in the earlier ice ages), the Near East, Russia, China, and Japan (by 15,000 B.C.). This culture also reached the New World.

Upper Paleolithic peoples were the first to live in the Arizona area, and they lived there for a very long time, though they made no lasting impact upon the environment. They were unspecialized hunters and food gatherers, capable of killing anything from rabbits to huge elephantlike animals. They made wide use of the skins of animals and lived in semisubterranean dwellings. Humans in the New World probably had this culture until about 10,000 B.C., when a change occurred; humans began to specialize in the killing of large animals.

THE BIG-GAME HUNTERS

Beginning about 10,000 B.C., there evolved in North America a way of life that specialized in the hunting of big game, and the people who lived this life are called the "big-game hunters." They undoubtedly continued to hunt small game and gather plant food, but though we know virtually nothing about those aspects of their life, tools that have been found and the remains of some kill sites enable us to know a great deal about their hunting of large animals. These people specialized in the killing of such animals as the mastodon, mammoth, musk-ox, giant beaver, grizzly bear, horse, camel, and especially the giant bison. Only a few thousand years after humans began to specialize in killing such big game, most of that game became extinct. Humans undoubtedly played a role in that extinction, and the greatest impact the big-game hunters had on the environment was the extermination of those big animals.

The culture of the big-game hunters was widespread, from New England to Arizona, and changed through time. The first big-game hunters belonged to the Llano culture, and the Naco and Lehner sites in southeastern Arizona verify that people of this culture were in the Arizona area by at least 9000 B.C. After the Llano culture came the Folsom culture, and after that the Plano culture possibly followed, though little is known of the last and many deny that it existed.

The economy of the big-game hunters was centered on hunting the large animals. It is also likely that they used vegetable foods, but no trace of this aspect of the culture survives. These people were exceptionally good stoneworkers. Projectile points found at their sites are all relatively similar; they are usually long with three sides, and the width is about one-third the length. A characteristic of these points is a channel removed from each side, extending toward the base. The Llano culture took out only a small groove, but by Folsom times, the carved grooves extended from the tip to the base (Figure 3.1). No one knows why these grooves were made, and only the big-game hunters made them. These projectile points were probably fastened to the ends of light spears, and the spears were thrown with an atlatl, a spear thrower that literally extends the thrower's arm about a foot, giving him much greater leverage. The bow and arrow were not introduced into the Arizona area until just before the time of Christ.

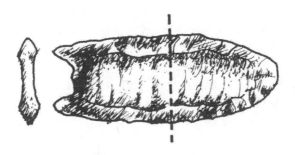

FIGURE 3.1. A typical projectile point used by the big-game hunters.

The favorite hunting technique was apparently the use of a "kill site," a place where animals could be driven over a small cliff, and those that survived the fall were killed with spears. Conservation was not a consideration, and many more animals were driven over cliffs than could have been used. The people made extensive use of animal hides, and they also made use of wood and bone tools.

THE DESERT CULTURE

Beginning about 6000 B.C., the big animals began to die off, and the way of life slowly changed as it became necessary to emphasize the small game and plant foods in the diet. This was a very slow change, and what resulted is called the Desert culture in the West and the Archaic culture in the East. In several parts of the Arizona area this way of life continued until long after the coming of the whites.

In many ways the Desert culture was an advance over the earlier big-game hunters since a great variety of tools and projectile points were produced. The fluted points were abandoned, and a great variety of stone points evolved as they were made for specific jobs; there were also many variations among regions and through time. The Desert culture also was an advance because the hunters and gatherers of that period took full advantage of the environment. They were still capable of killing big game, but they also took small game like birds and rabbits, and they had a greater dependence on vegetable foods. Their way of life was also more advanced in the sense that it was more sedentary. These people migrated, but they returned to the same sites year after year, and near those sites large trash dumps, or middens, developed.

The material culture of these people was quite complex. Whereas before this period there were only a few highly specialized tools, the Desert culture had bone fishhooks, bone awls, antler-tip projectile points, and other similar tools. The common projectile points had no groove, but they often had a stem at the base. The atlatl remained important throughout this period and in some areas even survived until the coming of the whites. Clothing was often made of rabbit skins that were cut into narrow strips and then wound and woven together to make quite good cloaks. Food was prepared once a day in a one-dish meal. The method of cooking consisted of digging a small hole, lining it with leather, placing water in it, adding hot stones to heat the water, and, when the water began to boil, adding the food gathered during the day.

The Desert culture began to come to an end about 1000 B.C., though in some places it survived until after the coming of the whites, and it was the introduction of pottery making and agriculture into present-day

United States that brought about the end of this culture. Pottery had been introduced into what is now the southeastern part of the United States from Mexico about 2400 B.C., but pottery found its way into the Southwest and present-day Arizona very late, only a little before the time of Christ. No one is sure of the reason for the delay, but possibly the Indians made such good use of gourds that they had no need for pottery. The history of agriculture is just the opposite. Maize had been introduced to the Southwest about 2500 B.C., but there was none in the Southeast until much later.

Toward the end of the Desert culture, many changes occurred in the Indian populations. There was considerable population growth, and regional adaptations began to be obvious. Trade also developed with a movement of goods over long distances. With the acceptance of agriculture, the beginnings of civilizations soon developed.

DEVELOPMENT OF AGRICULTURE IN THE NEW WORLD

Agriculture has a long history in Mexico. Maize was a staple by roughly 5500 B.C., and there is the probability that beans and squash were domesticated at an even earlier date. The origin of maize is unknown, but teosintle (or teosinte), a grass with a large head, seems to be in its ancestry. Maize may have evolved from teosintle, or perhaps that grass was crossed with a wild maize, which has since died out, to produce the domestic maize. The earliest maize was about the size of wheat; by 2500 B.C., cobs were about 3 in. long (7.6 cm.), and that size persisted for well over 4,000 years until the mid-1800s, when hybrid corn began to be developed. Today, if one enters many of the old Indian ruins, such small cobs are often seen since corn was the staple of many American Indians.

Some crops were domesticated in what is now the United States, but there is no evidence that any were domesticated earlier than the introduction of other crops. In other words, after the local Indians learned how to plant and grow the introduced crops, they began growing and experimenting with the local plants. Some American plants that were domesticated were goosefoot, marsh elder, sunflower, pigweed, and scarlet runner. The Indians, however, became primarily dependent on maize after it was introduced from Mexico, and after the whites came to the United States, most of the local domesticated plants were no longer used for food.

Those Indian groups who accepted agriculture had their entire way of life strongly affected, because with agriculture, they were not dependent on the whims of mother nature to produce vegetable food. Now the Indians could plant their food, and this accomplishment resulted in a larger and more dependable food supply. The Indians now had more free time to develop other aspects of life such as arts and crafts and religion. They also became more sedentary, because their food was produced and stored in one area, and they tended to live in permanent structures. Also, the population increased to much greater numbers because of the larger and more dependable food supply. The result was a change in the structure of society, because the old tribal structure of the earlier hunters and gatherers, which was based on family relationships, no longer worked when there were more people. What then developed were societies or civilizations based not on family relationships but upon law and order, with some people much higher in the social structure than others. These Indian civilizations developed in two areas of what is now the United States, the Mississippi Valley and the Southwest, particularly Arizona. The Mississippi Valley civilizations, in this work called the Woodland, developed along the valley of the Mississippi and its tributaries. Within the Woodland, there were many cultures over several centuries, such as the Adena, Hopwell, and Mississippian. The southwestern cultures are called the Classic cultures, and this term particularly refers to the Anasazi, the Mogollon, and the Hohokam.

CLASSIC CIVILIZATIONS IN THE SOUTHWEST

Three major Classic cultures developed in Arizona: (1) the Anasazi, who lived on the Colorado Plateau and whose descendants are the Hopi; (2) the Mogollon, who are ancestors to the Zuni and lived in the area of the Arizona and New Mexico border; and (3) the Hohokam, ancestors of the Pima and Papago, who lived in the Salt and Gila River valleys (Map 3.1). A minor culture, the Patayan (or Hakataya), lived along the lower Colorado, and their descendants today are the Yuman speakers.

The civilizations in Arizona seem to have developed very slowly as the local Indians living in the Desert culture began to learn about agriculture. Gradually their population grew, and societies evolved, and eventually recognized civilizations developed. These civilizations were therefore not the result of Indians from Mexico moving north, pushing aside the local Indians and bringing a civilization with them. The only possible exception to this rule were the Hohokam, who possibly did move north from Mexico.

Since these cultures began slowly and were based on different Desert cultures, they were initially very different. Only in some ways were they similar; they all grew maize, beans, and squash, and they all made excellent pottery. The pottery styles and decoration, however, were very different. But by about A.D. 1100, these civilizations had become quite similar. There was apparently a great deal of trade and contact between these peoples, and with this contact the groups grew more and more alike.

The practice of agriculture was critical to these civilizations. The area was not really suitable for agriculture—the Hohokam lived in a very hot desert, the Anasazi lived in a cold, high desert with virtually no permanent streams, and the Mogollon lived in a mountainous area—but these Classic or high

HIGH CULTURES OF THE SOUTHWEST

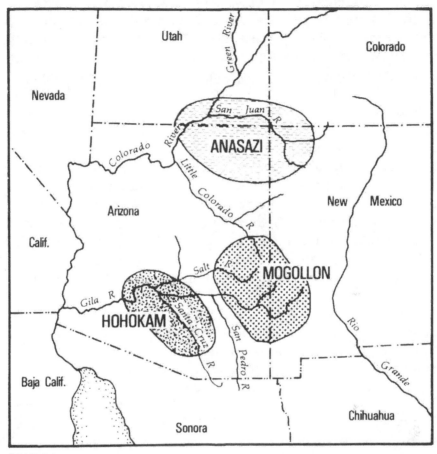

MAP 3.1

cultures would never have developed without agriculture. A major hindrance to these civilizations was the small number of domestic animals, the only ones being the dog, which was not eaten, and possibly the turkey. Because they lacked more domestic animals, a diversified village economy was not possible, and they were forced to continue hunting. But their populations were large, and they lived in harsh environments in which there was little game, so their diets suffered.

The Anasazi

The Anasazi culture developed on the Colorado Plateau, and its best-known ruins are in the Four Corners (for example, Mesa Verde in Colorado and Keet Seel in Arizona). This culture evolved from a local Desert culture called the Basket Maker culture. The Basket Makers began to practice agriculture and make pottery about A.D. 300, and that is considered to be the beginnings of the Anasazi. It is quite possible they initially learned these practices from more-eastern Indians, possibly from the Caddo Indians who lived in the Texas-Louisiana-Arkansas area, for their pottery and corn types initially were similar to those from that area. Subsequent additions to the culture, however, came from Mexico, mostly by way of the Mogollon Indians.

There was an orderly and evolutionary trend in the Anasazi culture as ideas were added or abandoned. After the idea of the bow and arrow was introduced, it was the dominant weapon by about A.D. 700, and the atlatl and light spear were slowly abandoned. About the same time turkeys began to be kept; cotton was introduced a short time later, and soon woven cotton blankets replaced fur robes. The potter's art also developed, and a wide range of vessel forms and decorative styles evolved; an especially popular style was black paint over a white base. Religion developed very early in the history of the Anasazi people, and soon religious activity was occupying much of their time. This religion changed only slowly through time and survives to this day.

One distinctive feature of the Anasazi culture was the style of the houses. The first houses were pit houses, low walls built around a shallow pit, which provided good insulation. The idea of a semisubterranean house was very common at that time in North America, and it may have originated in Asia, for pit houses were used at least 25,000 years ago in much of Asia. The house had a central fireplace; smoke exited through a hole in the center of the roof, and, with the use of a ladder, this hole was also the entrance to the house. Sometimes, however, an antechamber or passageway was used instead. The storage houses were almost identical: they were in a pit, circular, and lined with stone slabs. Only the absence of a fire pit would identify the structure to an archaeologist as a storage house, and some survived until early modern times.

The style of the Anasazi houses slowly changed. They began to be placed adjacent to one another, but circular houses did not easily lend themselves to such an arrangement. Therefore, it was not long before square houses were built, and a common wall was used by two houses. Putting together many of these square rooms resulted in the *pueblo* (the Spanish word for "town"). A pueblo was a very compact village, usually several stories high and it resembled an apartment building of today. The pueblos were built in different shapes such as L or U. All doorways opened toward the interior—for example, to the inside of the U—and walls enclosed all open sides. It then appeared to be a compact, multistoried building with no doors or windows opening to the outside. Some of the pueblos were quite large; for example, Pueblo Bonito in northwest New Mexico covered more than 3 acres (1.2 ha.), some of it was five stories high, it had over 800 rooms, and at its peak probably more than 1,000 people lived in it. At first, pueblos were usually built in the open, but toward the end of the Anasazi period they were mostly built in the large caves or overhangs that are common in the Four Corners area. It was these later, protected structures that survived to the coming of the whites, and many, such as Keet Seel, Mesa Verde, and Montezuma Castle in Arizona, are tourist attractions today.

The use of a circular pit house did not die out completely because it evolved into a ceremonial room known as the kiva. The kiva became a sub-terranean, round, sacred room in which a variety of religious functions were performed. It was used only by men, and much of their free time was spent in the kiva. Rainfall was uncertain and crop failures common, and, thus, most of the religious activity in the kivas was probably devoted to encouraging the gods to provide more rainfall. Each of the pueblos had several kivas within it, and most of the kivas would be in the open space in front of the homes (Figure 3.2).

The Anasazi were mostly known for their masonry construction, although it was quite rough masonry work when compared to that found in Europe where stones were cut and chiseled to shape. In the Southwest, rocks and flat slabs of stone were laid one on another, mud was used to fill the cracks,

FIGURE 3.2. Looking down on a portion of Mesa Verde in southwestern Colorado. The circular features are kivas with the roofs missing.

and often the walls were plastered. Various other masonry techniques and styles were developed by the Anasazi, and masonry construction eventually spread to many other groups in the Southwest. Another construction technique was called jacal. Large logs were set vertically in the ground, touching one another, the cracks between the logs were filled with mud, and often the walls were plastered. Yet another construction technique was wattle and daub. Wattle and daub was used to construct houses in the early phases of the Anasazi culture, and later it was used to divide rooms or to build temporary shelters. Wattle-and-daub construction consisted of large, widely spaced posts placed in the ground, and the rest of the walls was made of small sticks and interwoven reeds. Mud was shoved between the sticks to fill the open spaces, and the walls were then plastered. The walls were thin, but if the posts were large enough, they could support considerable weight.

The economy of the Anasazi, which in the Basket Maker stage had depended largely on hunting and gathering, slowly changed. The bow and arrow was the major hunting tool by A.D. 700, but there was little game left to hunt. Hunting had to be largely abandoned, and the Anasazi were then almost completely dependent on their agriculture. Their irrigation method was both ingenious and primitive. There were no permanent streams except the San Juan or the Colorado, and those were too big for the Anasazi to control, so they used a type of irrigation called flood irrigation. This method did not involve the use of canals to transport water; dams were simply built in convenient spots along a dry wash or near where a wash exited a plateau. A crop was then planted in the dry soil behind the dam, and the Indians prayed for rain. Sometimes they planted as soon as possible after a rain, and the water from that rain might be the only moisture the crop would get. The crops were not planted in neat rows but in little scattered hills. The corn planted on the outside of a hill would be dried up by the winds and did not produce a good crop, but that on the inside did bear well (Figure 3.3).

This was a very precarious type of agriculture, because several things could destroy the crop. The plants might die because of a second flood; a flood might be too large for the dam and simply wash it away; a severe flood could wash away the plants and soil; excessive silting could destroy some of the crop; there might not be any rain, or there might be too little to be of much use; and any number of other disasters were possible.

The Anasazi also planted crops in other ways. Sometimes terraces were cut into the hillsides, and since the terraces tended to conserve water, they provided a place where crops could be planted. Crops were also planted at or near a spring, and it was also common to find crops growing at the foot of a sand dune, because the surrounding sandy area would collect and conserve water and "transport" it to the foot of the dune.

The desert provides few places for this type of agriculture, and fields that resembled gardens were very widely scattered. The Anasazi farmers had to walk long distances from their central dwellings—at the time of the Euro-

FIGURE 3.3. Corn planted on the Hopi Reservation using the traditional, bunched method.

pean contact, it was not unusual for farmers to walk 20 mi. (32 km.) to their fields. Only rarely would the farmers live near the fields for protracted periods, and then only in order to try to protect the crops from birds and wild animals.

The decline of the Anasazi was swift, and apparently a major drought played a large role in the collapse of this society. The year 1276 was a very dry one on the Plateau, as was the next year, and the next, and many more—the drought lasted until 1299. Through the study of tree rings (dendrochronology), it has been determined that the drought was the severest suffered in the Southwest in the past 2,000 years.

The drought was a major reason why the Anasazi abandoned much of their territory. They had spread outward to areas that were, at best, only marginally suitable for agriculture. The drought made such areas unproductive, and they were abandoned as the people retreated inward toward the core, toward the slightly better agricultural land. This movement inward probably put an unbearable strain on the core area, for it meant there were many more mouths to be fed in the middle of a severe drought. It appears that this problem must have caused a time of unrest and internal conflict because many of the pueblos were destroyed and burned during this period. However, another possible factor was that the Southwest was invaded about the same time by a very warlike people, the Athabascans, the ancestors of the Apache and the Navajo.

The exact reason for the collapse of the Anasazi culture is not known, but it was probably the result of several factors, with the greatest being the drought and the subsequent internal problems faced by those peoples. The invasion of warlike peoples may also have played a role in the collapse, as such an invasion would have placed even greater stress on a civilization already in turmoil.

The Mogollon

The Mogollon lived in the mountains along what is now the New Mexico–Arizona border. Their descendants, the Zuni, still live there, and their reservation boundary abuts the Arizona border north and east of St. Johns. The Mogollon slowly evolved out of the Cochise variant of the Desert culture, and they were the earliest of the higher cultures in the Southwest, beginning about 300 B.C. They were, therefore, probably an agent in the transfer of ideas from Mexico to other groups in the area.

It is hard to characterize the houses of the Mogollon, because there was little architectural uniformity throughout their history. Usually the houses had their base in shallow pits and were one-room homes, a style that continued until the coming of the whites. The houses were constructed of large logs, sticks, branches, and reeds covered with mud. There were no real walls, but the roof, supported by posts within the house, slanted outward and rested on the ground outside the pit's edge. The floor was uneven because pits were dug in the floor for storage. Most houses had a roofed entryway coming in from the side, especially those built during the later periods.

The village form was very different from that of the Anasazi. The villages were a series of single-family dwelling units widely spread out along a ridge overlooking the cultivated land. Very few homes were located together, and apparently there was no scheme to the village as it appears that the houses were scattered at random along a ridge. There were deeply sunk or completely subterranean ceremonial structures in each of the villages. With time, the villages grew larger, and some were built on the level lowlands. Some of the village sites were occupied for over 1,000 years, which indicates the stability of this culture.

Many Anasazi ideas reached the Mogollon, and some were accepted. Masonry was introduced, and by A.D. 1100 it was widely used in construction. However, the Mogollon did not accept the Anasazi idea of living in compact apartment-style houses, and they continued to build the single-family isolated house.

The economy of the Mogollon was similar to that of the Anasazi. They had the same crops, and their major farming technique was flooding the fields. Hunting, however, always remained important to the Mogollon, as did the gathering of wild foods. The Mogollon always excelled at making pottery, and the first pottery in the region marked the beginning of their culture. At first their pottery was plain, either brown or red, and polished. It changed greatly over time, and the changes were mostly the result of internal innovation rather than ideas introduced from other cultures. The Mogollon culture, like that of the Anasazi, began to decline in the late thirteenth century. This decline was also probably related to the drought conditions, and possibly to invaders.

The Hohokam

The Hohokam civilization flourished in south-central Arizona, especially along the Salt and Gila rivers. Some authorities believe this culture evolved out of the Desert culture, but more and more evidence points to a northward movement of Indians living in Mexico. Perhaps it was a combination or mixture of those "Mexican" Indians and local Indians of the Cochise variant of the Desert Culture that led to the development of the Hohokam.

The Hohokam received a large number of their cultural traits from Mexico. If today's political boundary between the United States and Mexico followed the Salt River, everyone would agree that the Hohokam did exhibit rather typical traits of the high Mexican culture. There was little difference between this area and the regions to the south, except, of course, that this area was on the frontier, a long way from the center of the Mexican culture, so many of the traits were not so "sophisticated" as those to the south. Because of today's political boundary, however, some people find it difficult to recognize that the Hohokam were primarily influenced by the Mexican civilization.

Most of the agricultural techniques, including the Hohokam method of irrigation, and most of their agricultural crops came from the south. Over 300 ball courts have been found in present-day Arizona, and although they are not so elaborate as those found in Mexico, they are similar, and probably the same game was played on them. Many balls have been found at archaeological sites, and one rubber ball was obviously imported from southern Mexico. Small, stepped pyramids have been found in former Hohokam villages, and although tiny when compared to those found in Mexico, they do indicate that the concept of the pyramid did extend into the Arizona area. Many art motifs used by the Indians in central Mexico have also been found in Hohokam villages, and another indication of contact are the many trade items found that were probably imported into the Hohokam area, such as jewelry (rings, necklaces, and bracelets) and slate mirrors.

The houses built by the Hohokam were not pit houses as such, but their houses were in pits. In excavating such a house, one finds a pit about a foot and a half deep. Inside the pit, there are postholes all around the outside edge, and there are four large postholes near the center. The smaller holes around the edge were where the posts for the walls were placed. Most of the houses were of wattle-and-daub construction, with small posts to support the roof. The intervening spaces were filled with small sticks and branches, and mud was used to fill the cracks and plaster the walls. The roofs were quite heavy, and the four larger center posts helped support the weight. The roofs were made of large logs with smaller branches and reeds placed over them, and all was then covered with mud. A fire pit was located near the end of a short entryway. The shape of the houses varied through time, but toward the end of the Hohokam period, the houses tended to be ovaloid or elliptical and the later houses tended to be smaller in size.

There seems to have been little order or symmetry in the typical Hohokam village. Each village was encircled by a mud wall, and the individual houses were scattered about the enclosed area. All houses were about the same size, regardless of the importance of the inhabitants. A ball court and a pyramidal structure were often found within the village.

The Hohokam built many structures out of dried mud. This was not true adobe but puddled adobe. True adobe, mud bricks made in forms and then sun dried or fired, was introduced by the Spanish. The puddled adobe used by the Hohokam was fairly firm mud that was laid down to form a wall about 2 ft. (61 cm.) thick and 2 ft. (61 cm.) high. After the mud had completely dried, another layer was placed on top. The wall was thus slowly built, layer upon layer, and the top was thinner than the base (Figure 3.4). Large structures of puddled adobe, like the one at Casa Grande National Monument, were very unusual, and they were probably the result of Anasazi influence or perhaps even settlement by northern Indians among the Hohokam late in their history. The typical home was always built of wattle and daub.

The greatest and most impressive accomplishment of the Hohokam was their vast network of irrigation canals (Map 3.2). The first Hohokam irrigation was possibly flood irrigation along the Gila River, where several floods each year provided moisture for the crops. By A.D. 500 the Hohokam had begun to build small dams and canals. The dams were mostly made of rock and brush, and they did not store water but only raised it a little so it could be diverted into the canals. By using these irrigation techniques, the Hohokam were able to expand into the Salt River Valley and along the lower Verde, and the culture flourished. Their canals were not small. One at Pueblo Grande in the Phoenix area was 6 ft (1.8 m.) deep and 30 ft. (9.1 m.) wide, and some canals are known to have extended for 30 mi. (48 km.).

The canals were constructed at a great price in human labor. There were no wheels, metal tools, or beasts of burden, and all the work was done by hand. Small stone hoes were the major tool, and they had to be held in the

FIGURE 3.4. The ancient ruin of Casa Grande, which is built of puddled adobe. Some of the cracks between the horizontal layers of mud can still be seen.

hand because none were hafted. Excavated dirt and rocks were carried away in baskets on the backs of the Indians. In places, the canals extended through caliche and stone, and excavation there must have been extremely troublesome. Mainly the rock would yield only to hammering by other rocks, but possibly it could be fractured by lighting fires and then throwing cold water on the heated rocks. In maintenance, the accumulation of silt was a constant problem.

There was definitely an evolutionary trend in the Hohokam canal construction. The earliest canals were shallow and wide, and they were the easiest to build. However, the Hohokam probably realized they were losing water to evaporation because, with time, the canals were dug deeper and narrower. Some were as large as 15 ft. (4.6 m.) deep and 10 to 50 ft. (3 to 15.2 m.) wide, certainly as big as many of the larger modern canals. Also, if the banks were built up by using the excavated material, some of the canals could easily have been 20 or 25 ft. (6.1 or 7.6 m.) deep. The sides and bottoms of the canals were lined with clay. This technique was apparently used to prevent a loss of water in areas where a canal cut through sand or gravel, and it is evidence of the Hohokam's desire to conserve such a valuable resource. Another interesting aspect of the Hohokam canals was their

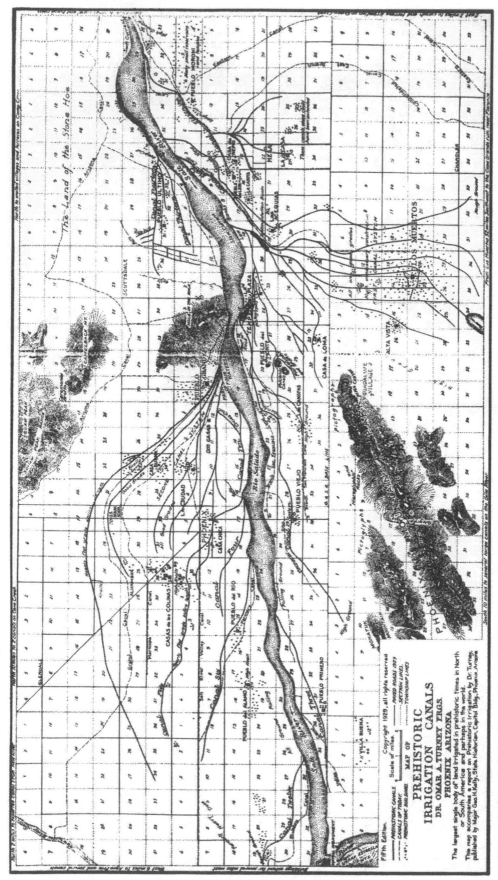

MAP 3.2. Prehistoric irrigation canals of the Salt River Valley. Courtesy, the Salt River Project.

length, because many were quite long, even longer than would seem necessary. The Hohokam probably had trouble with the soil becoming salinized and waterlogged, and their only solution was to abandon such land and to extend a canal into a new area. If this were true, and there is no reason to believe otherwise, the great length of the canals may have played a role in the decline of the Hohokam. Through time, the canals would have become even harder to maintain because of their increased length, and the work involved would have become too great.

The magnitude of the Hohokam canal projects tells us a fair amount about the people who built the canals. The Hohokam must have had a well-organized and controlled society, which means it must have been one that was stratified. The idea of castes was a common concept among the high cultures in Mexico, and there was probably an elite group in the Hohokam culture. A great deal of effort had to have gone into planning the canals, and great numbers of people were required for their construction and maintenance. Centralization of authority was required in controlling the labor, distributing the water, and bringing the water to several settlements. Activities had to be coordinated for the common good, and there was no place for several competing groups.

There is, however, little actual evidence that there was a ruling class among the Hohokam, although one must have existed. In Egypt and Central America, elaborate tombs provide evidence of ruling classes, but the Hohokam cremated their dead and had no use for such tombs. Even if they had built tombs, they probably would not have been elaborate ones. The desert provided only a marginal existence, and there was little surplus. Also, there were fewer people, so a ruling class would not have been able to acquire true power or a great many material things. It seems that the great amount of labor was for the good of all, not for the particular enrichment of a few.

Toward the peak years of these three high cultures, beginning about A.D. 1000, there was a great deal of interaction and contact among them. With time, traits that evolved in one area would be found in another. There was not only a movement of ideas and trade items back and forth, but also a movement of people. The Hohokam, for example, occupied the upper Verde Valley, but by A.D. 1100 Anasazi-like people, called the Sinagua, had moved south, and they lived there too. Another group from the north, the Salado, also moved to the Salt River area. The Hohokam and these new peoples appear to have lived in peace, and the Hohokam culture was somewhat changed as a result of the new ideas introduced by the new inhabitants.

By the 1300s all three of the high cultures were in decline. Possible reasons as to why this or that civilization declined can be guessed at, but why all three should have declined at the same time is certainly unknown. These civilizations did affect later peoples, however, for they are probably the source for the stories of the cities rich in gold, the Seven Cities of Cibola.

By the time the Spanish entered the area, little was left, and the Indians the Spanish found living there, probably direct descendants of those who lived in the high civilizations, had little to tie them to the earlier cultures. The Pima, for example, called the ancient civilization in their area the Hohokam, meaning "those who have gone before." When Father Kino entered the area, he went to marvel at and speculate about the Casa Grande ruins, just like a tourist today.

Other Indians

Another group of Indians that advanced toward the "beginnings" of a civilization was the Patayan (or Hakataya) along the lower Colorado River. But this group never progressed very far, and it should not be compared with the three high cultures. Patayan pottery, for example, never progressed much beyond planeware, and it remained simple in form. A probable reason why the culture of these Indians never really developed like the cultures of the others is that the Colorado River was just too big for them to handle and control. Concrete and steel dams were necessary before agriculture could be a true success along the lower Colorado.

It must be remembered that prior to the coming of the whites, many of the Indian groups in the Arizona area had not learned about agriculture, or at least had never accepted it, and thus they were still living in the Desert culture stage. Those Indians lived in what is now northwest Arizona and the central Mountain District. Some, like the Yavapai, engaged in a little agriculture to supplant their hunting and gathering way of life, but agriculture did not play a major role in their lives. Also, new peoples, ancestors of the Navajo and the Apache, were entering the area, and they largely rejected the sedentary way of life.

INDIANS OF THE SOUTHWEST AT THE TIME OF EUROPEAN CONTACT

When the first whites began to enter the Southwest – the Spanish from the south and then the Americans from the east – the Indians were living in a dispersed manner, and there was little or no cooperation even between subgroups. The Indians spoke many languages, had no high civilization, and were still living in what is termed the Stone Age. Thus, they were unprepared and unable to resist the new invaders. The impact the whites had upon these local Indians began in the 1540s, but some Indian groups avoided any strong white influence until the mid-1800s. The Southwest at that time can be divided into broad categories based upon the areas they occupied: Interior Basin and Range; California; Plains; and Southwest, the last occupying large parts of what is now Arizona and New Mexico (Map 3.3).

INDIAN ECONOMIC AREAS
AT THE TIME OF EUROPEAN CONTACT

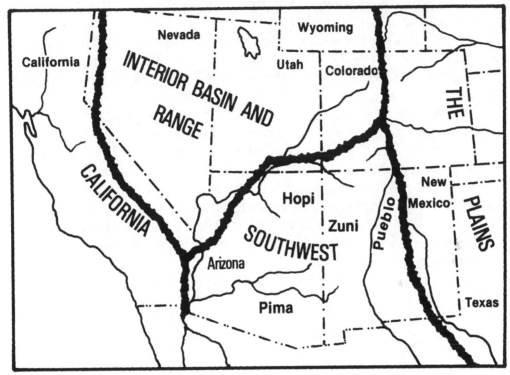

MAP 3.3

The Interior Basin and Range Indians, centered in western Utah and Nevada, were some of the most backward Indians in all of North America. The area they lived in can be thought of as a "cultural sink" because all the Indians around them were much more advanced—for example, the agriculturists to the south, the salmon fishermen in the Northwest, and even the Plains Indians to the east. These backward Interior Basin and Range Indians copied ideas from the surrounding peoples; for example, their traditional home was semisubterranean, but many to the east adopted the tepee common to the Plains Indians, and others, the longhouse typical of the Northwest Indians.

None of these Basin and Range Indians practiced agriculture, so they were hunters and gatherers. A typical example would be the Shoshoni, who were largely dependent on gathering plant foods and hunting rabbits so they had to move frequently. They made only a little pottery—although their basketry was well developed—and their tribal organization was very weak. They operated in bands of just a few families each, and each band was independent and had its own leader. This arrangement was typical among hunters and gatherers worldwide, and it exemplified the old Desert culture of the Southwest prior to the introduction of agriculture. This way of life spilled over into the Strip region of present-day Arizona, for the Ute Indians of that area were definitely hunters and gatherers.

The Indians in California (their area approximated the present boundaries of that state) were quite different, although their tribal structure, too, was very weak or nonexistent. There was no agriculture in all of the area except among the Yuman speakers along the lower Colorado. But food was plentiful, and the California Indians were largely dependent on acorns. The acorns were gathered in very large baskets, and it was a long process to remove the tannic acid before they could be eaten. Because of the availability of food, the population was dense, but it was also highly fragmented. There were many different tribes in California, and most of them spoke a distinct or a unique language—about one-third to one-half of the languages spoken in all of North America were spoken in the California region. This fact indicates there were many small groups with little or no mobility or interaction among them.

The Indians of the Great Plains were different. They largely hunted the buffalo for a living, and raiding and warfare were important in their culture. These people developed good tribal organizations and had chiefs. There was also cooperation and communication among groups—a late example might be the fact that several thousand men gathered to fight General Custer. At first agriculture was commonly practiced along the streams, but after the introduction of the horse, agriculture was largely abandoned as the Plains Indians began to specialize in hunting the buffalo. The Plains Indian traits were introduced into the Arizona area by the Navajo and the Apache.

The Southwest Indians, in the area of Arizona and New Mexico, cannot be categorized because there was great diversity in their life-styles. By 1540 when the Spanish first entered the area, many of these Indians still made their living as agriculturists, and the major agricultural groups were the Hopi in northern Arizona, the Pima in southern Arizona, the Zuni along the Arizona–New Mexico border, and the Pueblo Indians who lived along the Rio Grande in New Mexico. The culture of the Pueblo Indians developed quite late, and it is probable that during the great drought some Anasazi moved to the Rio Grande to take advantage of that dependable stream and that they were the stimulus for the development of the Pueblo culture. The Spanish were always attracted to these sedentary Indians because they had fairly large populations, lived in permanent houses, and usually had an agricultural surplus.

There were also many nonagricultural Indians in present-day Arizona. These Indians were always migrating in search of food, accumulated no wealth, and had small populations, and, thus, they were largely ignored by the Spanish. Other Indians knew about agriculture but were not dependent on it, and they only used it to supplement their hunting and gathering diet. These Indians, like the Papago and some of the Yavapai, planted crops in some areas, but they wandered as they gathered plant food and hunted, and returned to the fields only for the harvest. They, too, did not have permanent villages or large populations, and they did not attract the attention of the Spanish.

The Navajo and Apache were also hunters and gatherers, and they were a major force in the Arizona area when the Spanish arrived. The Apache and Navajo began entering the Southwest about A.D. 1100, but the exact date is unknown. Some authorities argue for an even earlier figure, and some argue that those Indians entered the Southwest as late as 1500, which is the approximate date of the oldest hogan so far dated in the Southwest. The Navajo and the Apache speak the Athabascan language, and the homeland of the Athabascan speakers is northwestern Canda and central Alaska, just south of the home of the Eskimo. Some splinter groups began moving southward, and eventually they developed the Apache and Navajo culture when they finally settled in the Southwest. They apparently migrated southward across the Great Plains and along the eastern foothills of the Rockies.

The culture of the Apache and the Navajo is a mixture of traits common to the Athabascan speakers of Canada and Alaska, the Plains Indians, and the Indians of the Southwest. The language has changed, and many Southwest Indian words have been added, but Athabascan speakers in Arizona can still recognize some words spoken by their cousins in Alaska and Canada. An example of a culture clinging to the old ways, and yet changing, is the Apache Indian girls' puberty rite. This rite of passage is still very important, and it is the principal element of religious activity for both the Athabascans in the North and the Athabascans in the Southwest. In the Southwest, however, many other Southwest Indian traits, such as the element of god impersonation, have crept into the ceremonies, including the girls' puberty rite. The Mountain Spirits (called *Gans*) are masked impersonators, and they are an example of god impersonation. This god impersonation is very strong in the Southwest and has entered the Apache culture. The traditional home of the Apache, the wickiup, and of the Navajo, the "forked-stick" hogan, have both been compared to the tepee of the Great Plains Indians, but they closely resemble the home of the Athabascans in Canada and Alaska. Some Plains Indian traits were adopted by the Apache and the Navajo, however, such as their early emphasis on raiding.

There was a constant conflict between the invading Athabascan speakers and the indigenous agriculturists. The Navajo and Apache were hunters and gatherers, and they often supplemented their diet by raiding the fields of the agricultural Indians. The Navajo and Hopi were soon enemies because both were trying to make a living in the same harsh environment, each in a different way, and conflicts were inevitable. Indeed, the two have strained relations to this day. The same was true with the Apache and their neighbors because the newcomers tried to fill ecological niches not filled by the local Indians. The Apache spent much of the winter in the mountains and summers in the valleys, the opposite of the local Indians. But the land was too harsh to allow both to earn a good living, and conflicts resulted.

INTRODUCTION OF THE HORSE AND ITS ROLE IN INDIAN CULTURE

The acquisition of horses greatly affected the lives of all the Indian groups. Some of the Indian groups, particularly the hunters and gatherers, used the horse to give them a great amount of mobility, but the agricultural Indians had little use for horses. Unfortunately, some of the hunters and gatherers, like the Apache, were warlike, and the use of horses turned them from a rather ineffective group of Indians into an effective mounted cavalry, all to the detriment of the agriculturists in the area.

The horse evolved in the New World, and before it became extinct there, about 7000 B.C., some horses migrated across the Bering Strait to the Old World. The Spanish brought the horse back to the New World at the time of their conquest of Mexico in 1521. The horses brought by the Spanish were small, sturdy, sure-footed horses of excellent stock that had some Arabian blood.

A very popular theory often believed by the layman and once advocated by scholars, was that the Indians first acquired their horses from wild herds. It was believed that some horses strayed from the expeditions led by the early conquistadores and that those strays thrived on the grasslands of America and produced large wild herds from which the Indians acquired horses. That theory is just not true. The conquistadores preferred to ride stallions, and therefore there were 588 horses on the Coronado expedition, but one authority states only 2 were mares. There were no mares at all on the de Soto expedition. The Spanish also considered horses to be very valuable – Cortez considered one horse worth 20 men – so the men always watched the horses carefully. If a horse were to stray, it was hunted and brought back. Also, when Oñate led his expedition across the Southwest, 60 years after Coronado, he observed game and the Indians and their use of dogs, but he made no mention of seeing any wild horses. This would not be an oversight. Also, the first recorded mention of a wild herd of horses on the Great Plains was not until 1705. Even if there were wild horses earlier, the Indians would have feared such animals and would not have known how to catch, tame, and control them. The Indians probably would have considered horses as something good to eat. When the de Soto expedition, for example, abandoned 6 horses on shore, they were all killed and were being butchered before the Spanish were out of sight. There was just no way Indians could have acquired horses from wild herds.

The Indians first acquired horses when they were given them by the Spanish, and then the use of horses spread from one Indian group to the next. The first recorded instance of Spaniards giving any horses to the Indians was in 1541 in Mexico when the Spanish and the Indians were fighting a common enemy. In 1582 horses were given to some Indians in central Mexico as a bribe to keep the peace, and by 1598 horses had been given to Indian guides and allies in the El Paso area. Soon all the Indians in the Southwest and northern Mexico had acquired horses.

It was against Spanish law to teach Indians how to use the horse, but the law was frequently broken. Missionaries, miners, ranchers, and military personnel all taught friendly Indians about horses, or at least gave the Indians a sense of familiarity with them. Many "friendly" Indians who had learned about horses went to live with "unfriendly" Indians. They may have taken stolen horses with them, or they may have raided Spanish settlements for them. Soon many Indians knew how to use horses, and those Indians taught other Indians.

The use of horses spread rapidly from one Indian group to the next. By 1550 there were mounted Indians in northern Mexico; by 1600 horses were commonly used in New Mexico, and some Indian groups even had herds of horses; and by 1700 there were horses all across the Great Plains, northward toward Montana and eastward to the Caddo territory in the Texas-Arkansas-Louisiana area. When La Salle first visited the Caddo area, the people there already had horses, which they called *cavalis*. The Plains Indians, therefore, had been using horses for over 100 years when the first Anglo Americans began to penetrate that area, and those Indians could resist the invaders as a result. The same was true with the Apache and the Navajo in the Arizona area.

The Indians mostly used Spanish techniques in relation to horses. They mounted on the right side, which is a Spanish (and Moorish) trait, and Indian armor (for horses and men), tack (bridles, saddles, etc.), riding styles, and even weapons were almost exact models of the Spanish equivalents. The Indians acquired some of these goods by trading and others in raids. Many Indians on the Great Plains used lances in combat, and others used lassos to rope their livestock.

One technique few Indians mastered was the breeding of horses. The Plains Indians understood horse breeding because before they acquired horses, they had bred very large dogs to pull the travois. Most Indians in the Southwest, however, had to constantly raid to acquire more horses. These raids soon became important to their culture because it became the way to prove one's manhood. After the Indians were forced to halt their raids, it usually took them many years to make any headway in increasing their own herds.

SPANISH EXPLORATION

Spanish settlement in the New World took place soon after the initial discovery of the area by the Europeans. Santo Domingo, on the island of Hispaniola, was established in 1496 (after having been established elsewhere first); Cuba was first settled in the early 1500s; and Florida, by 1566. The effect of the Spanish upon the natives was disastrous. They were rounded up, made slaves, and forced to work in the mines and in the fields. Because of disease and slavery, their numbers decreased at an alarming rate. The Spanish sent out many slaving expeditions and virtually depopulated the islands except for the very warlike Caribs in the extreme south. Soon the Spanish were importing Negro slaves.

Expansion westward began in 1517 after rumors began circulating in Cuba of immense wealth to the west. An exploratory expedition was sent out, and in the area of Vera Cruz, the explorers acquired a fortune for just a few trinkets. The rumors were confirmed, and this success led to the conquest of Mexico by Cortez in 1521. The result was immediate wealth, and thereafter, for about 100 years, any rumor of gold or wealth sent the Spaniards scurrying off in search of wealth, power, and fame.

The Spanish used the Mexico City area as a base of operations for their explorations. There were thousands of Indians in that area and much surplus food, and the Spanish used both to good advantage in outfitting expeditions. Exploring and conquering expeditions were soon sent in all directions. The Spanish considered some of the expeditions very successful, because of the wealth produced, but others, such as those that extended northward into the present-day United States, were considered failures.

The expeditions to the north were organized because of reports of gold and of cities with large Indian populations in that direction, and the expeditions were sent out to check on those rumors. The expeditions to the north did not locate either great wealth or large numbers of Indians, and, therefore, there was not immediate Spanish expansion and settlement in what is now the United States. It was to be another 60 years before the Spanish began to settle in the southwestern part of present-day United States, and even then their settlement was along the Rio Grande. From the Rio Grande, Spanish influence spread to the Indians in what is now eastern Arizona, but it was not until almost a century and a half after Coronado that Spanish occupation, through the mission system, began to extend northward into the Arizona area.

Cabeza de Vaca

The first Spanish explorer into the Southwest, and possibly the first European to set foot in Arizona, was Alvar Núñez Cabeza de Vaca, and his was far and away the most interesting and unusual of all the expeditions in the

New World. The title Cabeza de Vaca ("head of a cow") had been given to the explorer's maternal great-grandfather after he guided the Christians into battle against the Moors, using the skull of a cow to mark the way, and the younger man always preferred to be known by that title. Cabeza de Vaca had been on several expeditions before he was made treasurer of an expedition led by Pánfilo de Narváez, who had been given a patent by Charles V, king of Spain, to subjugate and colonize the entire Gulf Coast. Narváez commanded the expedition, which consisted of 600 men, many horses, and five ships, when it set out from Spain in 1527. The expedition was plagued with bad luck from the beginning. The trip across the Atlantic was a rough and hard one, and in the West Indies several men deserted and there were hurricanes. Almost a year after leaving Spain, the expedition finally landed in Florida, near Tampa Bay.

Unfortunately Narváez divided his men, and he and about 300 others headed into the interior – and trouble. The Spanish suffered constantly from hunger, Indian attacks, and disease. In northern Florida, the Spanish finally gave up and headed to the coast, but they failed to link up with the waiting ships. The Spanish then constructed five boats of horsehide and made the decision to sail westward, because they thought Mexico was much closer than it actually was – rescue would have been six times closer if they had gone the other way. There were 45 men in the boat headed by Cabeza de Vaca. That boat and another one were finally washed ashore and wrecked on Galveston Island. Many Spaniards died, and soon only a small handful were left. Cabeza de Vaca made his living as a trader, and he traveled freely between native villages. He also developed a reputation as a healer and medicine man. After several years, four of the shipwrecked sailors decided to leave and travel westward (Map 3.4).

The four men who traveled west were Cabeza de Vaca, two other Spaniards, and a black slave named Estevanico. They were thought to be healers and medicine men, and their reputations allowed them to travel westward. They eventually reached the Rio Grande in the area of El Paso, but the route they took from that point is uncertain. Some authorities argue that they headed due southwest, in which case they missed Arizona, but it is quite likely that they went northward and then turned westward and southward to perhaps travel through the very southeastern part of what is now Arizona in the year 1536. They finally met some Spaniards who were on a slave-gathering expedition along the Sinaloa River. Their trip ended almost exactly eight years from the day they landed in Florida.

Cabeza de Vaca was welcomed back to "civilization" and considered a hero. He returned to Spain and was eventually given a high position in Paraguay. He was not well received there because of political problems (probably because of his feelings about brutality and the enslaving of Indians), and he spent a considerable amount of time in jail and was banished

PROBABLE ROUTE OF CABEZA DE VACA

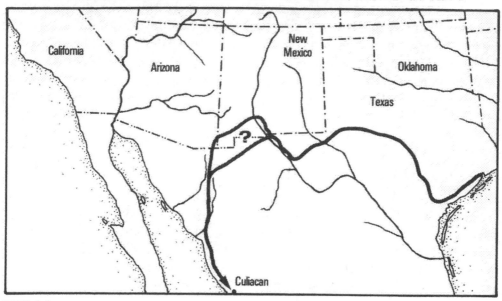

MAP 3.4

to Africa before dying peacefully in Spain. The other two Spaniards quickly passed from history, but Estevanico was less fortunate. He was still a slave and had to remain in the employ of the Spanish, and he had one more important role to play.

Although Cabeza de Vaca barely entered the area that is now the state of Arizona, if at all, and returned with no gold, his impact was great both in regard to stimulating interest in exploring the area and in regard to changing the attitudes of the Spanish toward the Indians. He had started the journey as a hardened conquistador, but he returned a tolerant person concerned about justice and equality. He insisted that the system of Indian slavery (*repartimientos*) must be ended, and he expressed his views in his work *Relation of Alvar Núñez Cabeza de Vaca* (trans. Buckingham Smith, [New York, 1871]).

> We passed through many territories and found them all vacant: their inhabitants wandered fleeing among the mountains, without daring to have houses or till the earth for fear of Christians. The sight was one of infinite pain to us, a land very fertile and beautiful, abounding in springs and streams, the hamlets deserted and burned, the people thin and weak, all fleeing or in concealment. . . . and they related how the Christians, at other times had come through the land destroying and burning the towns, carrying away half the

men, and all the women and boys, while those who had been able to escape were wandering about fugitives. . . . Thence it may at once be seen, that to bring all these people to be Christians and to the obedience of the Imperial Majesty, they must be won by kindness, which is a way certain, and no other is. [Pp. 174–75]

Cabeza de Vaca's account was widely read, and the Spanish at home were appalled at what had been happening. The work soon led to a complete change in policy toward the Indians, and afterward the Spanish began to convert the Indians to Christianity rather than to make slaves of them.

Cabeza de Vaca also influenced the Spanish in other ways. He reported that there were gold and large tribes to the north, though he had never met them and had only heard of them through rumors, and this information was to lead to the Marcos de Niza and Coronado expeditions. He described many animals never seen before by the Spanish, such as buffalo, opossum, and armadillo. He also influenced the mapmakers because they had never before realized the great width of the New World. He described many tribes and customs, and since he lived like an Indian and was accepted by them, his reports were accurate. The Spanish found it hard to believe some aspects of his report, such as the fact that homosexuality was common and accepted in the societies, but in the end he was proved correct.

Expedition of Marcos de Niza

Cabeza de Vaca's report led directly to the exploring expedition of Marcos de Niza in 1539. Cabeza de Vaca had not seen the rich cities or great amounts of gold, he had only heard tales from the Indians of rich cities to the north, but still his report excited the interest of the Spaniards in Mexico City. The Spanish were probably hearing what they wanted to hear. The Indians always told the Spanish what they thought they wanted to hear, and the Spanish probably asked questions in such a way as to get the answer they wanted—for example, "Have you heard of 'rich cities'?" At any rate, the Spanish believed there were seven rich cities to the north, the Seven Cities of Cibola. The story of seven rich cities was an old one to the Spanish, and it probably originated in a tale of how an archbishop and six bishops fled westward from the Moors in seven ships and how they eventually established seven wealthy Christian cities. Cabeza de Vaca's reports seemed to reaffirm the old tale, even though he probably only heard local Indians describe the Hopi, Zuni, and possibly the Pueblo Indians.

The Spanish quickly decided to check on these stories of rich cities and gold and looked for someone to lead an expedition. Cabeza de Vaca and the two other Spaniards who had been on his trip refused so the guide selected was the unfortunate slave, Estevanico—he had been purchased from one of the surviving Spaniards by the viceroy of New Spain (Mexico) and had no choice in the matter. The expedition leader was Fray Marcos de Niza, a Franciscan who was born in what is now France. He had been on expeditions in South and Central America and therefore had experience in such matters, and he could be trusted.

The trip did not go well. Near the present border of Arizona, Estevanico was sent ahead, and Fray Marcos followed, but the two were never to meet again. Fray Marcos was following Estevanico when he received word that the black, and some friendly Indians, had been killed by Indians at Cibola. Fray Marcos claimed to have advanced to where he overlooked the first and smallest of the Seven Cities of Cibola (a Zuni village), and he thought it was a rich city even "bigger than the city of Mexico." He then beat a hasty retreat to Mexico City.

Marcos de Niza's report contained many basic errors. He was obviously wrong when he said that he saw, with his own eyes, that Cibola was larger and richer than Mexico City; he also claimed that he was never far from the sea and that another expedition could be supplied by ship. These mistakes were probably not lies but rather errors in judgment and interpretation. Much of de Niza's information came from the Indians, and they may have misled him or perhaps he did not understand them. Other mistakes, such as mistaking a Zuni village for a city the size of Mexico City, were definitely his. Some people have even gone so far as to come to the conclusion that he never even saw a Zuni village, except in his imagination, and that he began his return immediately after he heard the news of Estevanico's death. Nevertheless, his report seemed to confirm the legend of the Seven Cities of Cibola, and the Spanish prepared for conquest.

The Coronado Expedition

The major exploring expedition into the Arizona area in the early years was the one led by Francisco Vásquez de Coronado. This expedition left Mexico City in 1540 intending to conquer and exploit the rich cities that were believed to exist to the north. It was, by Spanish standards, a large expedition, consisting of approximately 300 Spaniards and 1,000 Indians, as well as some 1,600 horses and mules and several thousand cattle and sheep to be used for food.

Throughout the expedition, Coronado and the fighting men were in front, and the friendly Indians and animals came behind. Most authorities believe the expedition entered the southeastern Arizona area, followed the San Pedro River northward, and then turned northeast and crossed the White Mountains (Map 3.5). At that point, Coronado and his men were tired and hungry, but they still eagerly looked forward to seeing Cibola. But what the Spanish termed Cibola were the Zuni villages, the first of which was named Hawikuh. The Spanish fought a short fight with the Zuni and then entered that first village. The Spanish were extremely disappointed to find no gold, no silver, no precious stones. The only things the Zuni had that the Spanish

EXPLORATION BY CORONADO AND HIS MEN

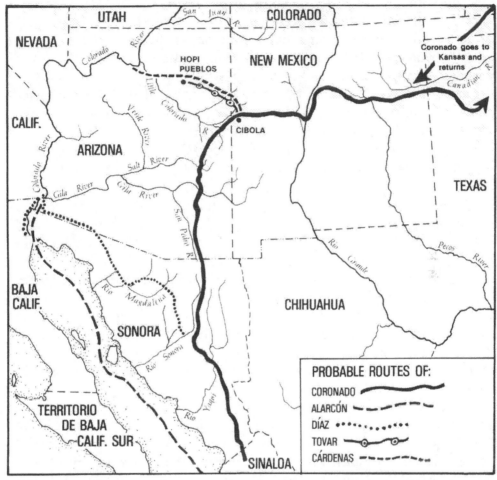

MAP 3.5

were happy to see were shelter and abundant food.

Marcos de Niza had been the guide for the Coronado expedition up to this point, but now he had completely lost his credibility. He was sent home in disgrace and had to take a letter from Coronado to Viceroy Antonio de Mendoza, that stated, among other things, "I can assure you that he [Fray Marcos] has not told the truth in a single thing he said, but everything is the opposite of what he related, except the name of the cities and the large stone houses" (Paul Horgan, *Conquistadores in North American History* [New York: Farrar, Straus and Co., 1963], p. 171). Fray Marcos de Niza, either justly or unjustly, has been known ever since as the "the lying Friar."

The dreams of riches were shattered, and Coronado had to reassess the goals and objectives of his expedition, with the realization that his previous actions had been based on fantasy. He questioned the Indians and sent out reconnaissance parties in order to obtain hard data upon which to base future decisions. One of those minor expeditions, led by Hernando de Alvarado, explored the Pueblo area of the Rio Grande Valley, and Alvarado advised Coronado to spend the winter there rather than at Cibola, and this he did. Hernando de Alvarado and his men continued east to the Pecos River, where they captured an Indian they named "the Turk." The Spaniards became excited about a place the Turk described called Quivira, a place literally overflowing with gold. Other Indian captives repeatedly told the Spanish that the Turk was lying, but the Spaniards insisted on believing his story – after the disappointment of Cibola, they were desperately grasping at any hope, although they should have realized that Quivira was also a myth.

The Spaniards spent the winter along the Rio Grande (putting down several small revolts by their hosts), and in the spring they set out to find and conquer Quivira. They reached central Texas before they recognized the stories of Quivira for what they were – myths. Coronado sent the main body of the army back to the winter quarters on the Rio Grande, and with another guide and a handpicked detachment capable of greater speed, he set out to see Quivira. Even though he now knew it must be a myth, he had to remove any possible doubt. Quivira proved to be only the grass huts of the Wichita Indians in central Kansas. After this discovery, there were no more attempts to discover gold. Coronado returned and again wintered among the Pueblo Indians, and then, using the same route, he returned to New Spain in 1542.

But the expedition did accomplish and discover a great deal, because Coronado sent out many reconnaissance parties. These minor expeditions ranged far and wide and accumulated much information. There were many of these small expeditions, but the four most important were led by Hernando de Alarcón, Melchior Díaz, Pedro de Tovar, and García López de Cárdenas (Map 3.5).

Hernando de Alarcón took three ships along the west coast of Mexico, supposedly in order to supply Coronado with needed supplies. Alarcón already knew the area, for he had explored the Gulf of California with Francisco de Ulloa in 1539. Alarcón took his ships to the mouth of the Colorado, but the ships could not cross the sandbars there, so Alarcón had two small boats loaded and ascended the river with 20 men. The current was so swift the men had to use ropes to pull the boats from shore. They soon met some Indians who, after a time, agreed to help pull the boats upstream. The Indians had heard both of Estevanico's death and of Coronado; but Alarcón found no Indian or Spaniard willing to try to link up with Coronado, so he eventually left messages in case some of Coronado's men came there, returned to his ships, and sailed for home. Alarcón was undoubtedly the first Spaniard to reach that part of the Arizona area. He claimed to have traveled 85 leagues up the river (about 311 mi. or 500 km.), but he did not mention the Gila River (only about 80 mi. [129 km.] from the coast), so it is not known exactly how far north he traveled.

Coronado realized that his route was carrying him away from the coast, but he still wanted help from that direction, so Melchior Díaz was assigned the task of attempting to contact the supply ships. Díaz was a capable man. He had been commander at Culiacán, the northernmost Spanish settlement, when Cabeza de Vaca returned to civilization. He had also led a small expedition into Arizona (before Coronado's) to check on Fray Marcos's account, and his report had been much more sober than that of the padre. Díaz's trip to the Colorado River, with some 25 Spaniards and some Indians and livestock, crossed a barren desert. For much of the way, their route was approximately the same as the present United States–Mexico border. When they reached the Colorado, Alarcón had already left, but Díaz found his messages.

Díaz explored the vicinity and traveled into present-day California, but he died an accidental death, impaling himself on his own lance. His men returned to Culiacán and sent a full report northward, but by then Coronado had given up all hope of resupply and was already in the Rio Grande Valley.

While at Cibola, Coronado was told of a rich area to the west, and Pedro de Tovar, with 23 men, was sent to check on the report. They found seven Hopi villages but no gold or silver. This was the first meeting between whites and Hopi, and there was a small fight – an indication of future relations. The only thing of interest that Tovar learned was that there was a great river to the west, "where people lived who possessed wealth." But Tovar was not authorized to go further, and he returned to Cibola.

Coronado sent García López de Cárdenas and a force of 25 horsemen to find and explore that river, and he hoped they might contact the supply ships. Cárdenas went to the Hopi villages and then westward to the Grand Canyon. He was the first white man to see the canyon, but he was not impressed. He wanted gold, and even more immediately he wanted water, and he was denied both. Volunteers tried to get down to the river's edge and failed, so after three fruitless days of searching for a way down, they returned and reported their reconnaissance had been a failure.

SPANISH SETTLEMENT IN NEW MEXICO

Spanish settlement moved rapidly northward from Mexico City along the Central Plateau of Mexico, and the Conchos Valley in northern Mexico was settled as early as 1570 by miners and missionaries. Two towns were established there, Santa Bárbara and San Bartolomé, and they soon became the bases from which the Spaniards advanced northward into present-day New Mexico and northern Arizona. By the early 1580s, there were again rumors of gold and Indians without Christianity to the north, and these

SETTLEMENT OF NEW MEXICO
AND ARIZONA EXPLORATION

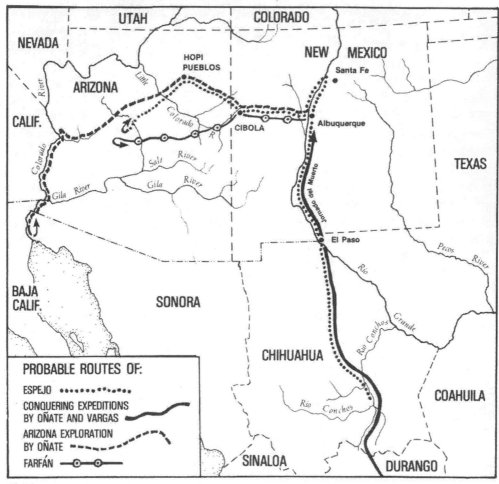

MAP 3.6

rumors fired the imagination of the Spaniards. Sir Francis Drake returned to England in 1580 after a profitable trip of thievery in the Pacific, and the Spaniards were concerned that maybe he had found the Northwest Passage. For all these reasons they were very interested in extending their control northward. Several expeditions went north, and one in particular traveled through much of what is now Arizona. This was the Espejo expedition of 1582 that was jointly led by Antonio de Espejo and Bernadino Baltran, a Franciscan (Map 3.6). They traveled into the Arizona area for riches, and apparently they were the first whites to discover the rich ores in the mountains near present-day Jerome.

Reports had been forwarded to Madrid, and a decision was made at the royal court to colonize the Pueblo area along the Rio Grande. The person who received the contract to conquer and colonize the area, at his own personal expense, was Juan de Oñate. He was a very wealthy man, an heir to a fortune in silver, and he was very powerful politically, since his wife was the granddaughter of Cortez and the great-granddaughter of Montezuma. Oñate moved north with many soldiers, families, missionaries, friendly Indians, and thousands of animals. He took possession of the Pueblo area in 1598, and within a few months the occupation was so successful that Oñate began to think of exploring for wealth. He sent out several expeditions into the Arizona area, and one, led by Marcos Farfán, reached the central part of the state. Oñate led some of the expeditions himself, and one of those was to central Kansas to visit Quivira–ideas of wealth die hard. In 1604–1605, he led another expedition through the Arizona area and went all the way to the mouth of the Colorado River (Map 3.6).

That expedition ended the great era of exploration in the Arizona area that had begun in 1540. By 1605, the area was fairly well known by the Spanish, and they were convinced it contained neither great wealth nor the Seven Cities of Cibola. Some rich ores had been discovered, but they were in the middle of a wilderness thousands of miles from Mexico City; besides, richer ores existed to the south. For the next 75 years, there was little Spanish interest in the Arizona area, except for some sporadic interest in the Hopi.

The earliest Spanish capitals in New Mexico (there were two) were not centrally located, being too far north and too near established pueblos to be effective capitals. Santa Fe was, therefore, established in 1610. It was a new town, centrally located in Pueblo country, and although it never grew to be a major city, it still functions as the state capital. Albuquerque, further south, quickly became the major city of the area in terms of population and commerce.

The Spanish began to exert some influence in the northeastern Arizona area from their base in the Rio Grande Valley. They never conceived of a political boundary between that area and present-day New Mexico, and they could see no reason to consider the Hopi or Zuni Indians as anything but other "Pueblo" Indians that were a bit isolated from the rest. The Spanish missionaries never went to work among the Navajo because the Navajo were hunters and gatherers living in scattered and temporary settlements, but the missionaries were very interested in the sedentary Hopi, and several Franciscan friars worked among them at various times. Initially they were very successful, but gradually resistance developed among the older, more traditional Hopi as they saw the old ways being replaced or destroyed. When the Pueblo revolt began in 1680, the Hopi joined with enthusiasm and killed the four missionaries living among them. Never again would the Hopi be greatly influenced by missionaries. After the Pueblo revolt, the Spaniards attempted missionary work among the Hopi, but only halfheartedly. The Hopi land was harsh and did not interest the Spanish, so those people were allowed to go their way and were largely ignored by the Spanish. Mormons, Catholics, and various Protestant groups have worked among them, but all with a limited amount of success. Today the Hopi largely cling to their own cultural ways, and they still maintain their own beliefs, religion, and ceremonies.

The Pueblo Indians of New Mexico revolted in 1680, and the uprising was a bloody one. As many as 2,500 Spaniards and friendly Indians fled to the El Paso area, and they remained there for over a decade. The reconquest of New Mexico was led by Diego de Vargas in 1692, and except for a brief revolt in 1696, the Pueblo area was again firmly in Spanish hands. Vargas went into the Arizona area in an attempt to subdue the Hopi, but there was no fighting as the Hopi swore allegiance to the king of Spain, an act that probably meant little or nothing to them. From then on, they were largely outside the realm of Spanish influence.

The Pueblo area was thus made a Spanish territory, but it was a very isolated one. The area was surrounded by hostile bands of aggressive wandering Indians: to the east, the Comanche; to the south, Apache; to the west, Navajo; and to the north, the Utes. By the mid-eighteenth century, those tribes had become dangerous foes, for they were mounted and greatly embittered by a long history of cruel treatment at the hands of the Spanish. This area of Spanish settlement was, therefore, like an island. The only link to the outside world was across the Mesilla Valley south to El Paso, but that route was so dangerous that it was called the *jornado del muerto* ("journey of death"). As a result, there was very little Spanish expansion outside the area, and the only "Spanish" intrusion into the Arizona area from this direction was very late and resulted in only a few settlements, such as St. Johns and Concho. That expansion occurred only after the American occupation and often was with the support and backing of Texas cattlemen. The next Spanish movement into the Arizona area was from Sonora to the south.

THE MISSIONARY PERIOD IN ARIZONA

When the Spaniards decided to control an area, they essentially used three ways to accomplish their goal: (1) establish a presidio; (2) establish a civil community, which could be a *ciudad* ("city"), a *pueblo* ("town"), or a *villa* ("village"); and (3) establish the mission system. They seldom used only one way; instead in occupying an area, they often used two or all three in their overall scheme of settlement. It was intended that all presidios and missions would eventually be converted into civil communities.

A presidio was a walled fort usually located at a strategic site suitable for grazing and a little agriculture. The commander was usually a captain; there was often another officer or two and from 20 to 50 enlisted men. They had various functions, but the most obvious was to protect the local Spanish settlers, missionaries, and friendly Indians from hostile nomadic Indians. The military would also, by their presence, help prevent rebellions.

A pueblo or villa was an economic, governmental, and administrative center, and it was more like an ordinary town or settlement. There were often farmers associated with the town, and many of them lived in the villa and commuted to their fields. A villa was not meant to serve as a fort or as a place to attract Indians. Los Angeles, Albuquerque, and Tucson are examples of present-day cities that began as villas.

A mission system was much more complex, and it was quite important for Arizona. The obvious objective of a mission system was the conversion of souls, but the mission system on the frontier performed many more functions than that. A missionary felt responsibility toward (1) the Roman Catholic Church, (2) the Spanish government, and (3) the local Indians.

A mission was a permanently settled site with a resident priest, but it is hard to describe the typical mission as no two were alike. There were, however, some characteristics common to all the missions. As in most Latin countries, the location was on high ground, so that workers in nearby fields could look up to the church, plus such locations were excellent for defensive purposes. All missions were established near streams to assure irrigation, and rainwater for use in the mission proper was usually funneled from roofs into underground cisterns. The entire mission had a high wall around it.

The most obvious and imposing structure in a mission would be the church, and such structures often took 10 to 15 years to complete. (Since these churches were substantial structures, many have thus survived even though all the other mission buildings and walls have fallen into ruin and disappeared.) Along the inside of the wall surrounding the mission would be the other buildings that made up the mission. They would have rooms for various purposes, including the living space for priests and the many Indians who lived within the mission compound. Other rooms would be used for such things as a rectory, storage, kitchen and dining facilities, granaries, schoolrooms, and shops such as carpenter or blacksmith shops.

Within the walls was a very large open area, called the patio, where all the Indians could assemble for lectures. The area within the walls was quite extensive and often included, among other things, a garden, a small graveyard, and a chapel or two. A small military detachment was not uncommon, and the mission often played the role of a fort in times of Indian conflict. In such times of danger, it was not unusual for many Indians to live within the mission walls and for animals to be kept within the patio at night. Outside the mission walls would be found an Indian village, cultivated fields, corrals, orchards, vineyards, and the like. Animals were herded and grazed on the open range around the mission, and the fields were usually fenced. Each mission was meant to be largely self-supporting, and it was the home base from which the padre could go out to work with Indians who lived far from the mission.

The mission system was a hierarchical one, and the major element was the mission with its resident priest. Next came *visitas,* which were visited regularly by the missionary, and finally, *rancherias,* which were visited only rarely. The first mission established by a missionary when he began his work had to be a success or all of the rest of his work in that area would fail. For about a year, a padre would concentrate his efforts on the first mission, the "mother" mission, trying to ensure its success. Once the mother mission was a success and functioning efficiently and well, the missionary would begin to work with other Indians by establishing sub-missions or *visitas.*

A *visita* was a small incipient mission established near an Indian settlement, and it would have a small chapel. The missionary would periodically travel to a *visita* to work with the Indians. He would teach about Christ, and at the same time he would help the Indians improve their agriculture and life-style by introducing new ideas, crops, and domestic animals. If many Indians moved to a *visita,* it was usually made into a regular mission with its own resident priest, who would then begin establishing more *visitas* out from the new mission. Many missions thus started out as successful *visitas.*

A *rancheria* was a small Indian settlement that a missionary would visit only infrequently. At a *rancheria,* a padre would have the Indians build a small chapel, or at least a ramada, a small roofed structure without walls, where Mass could be held and where the Indians could be taught. There was often some agriculture at a *rancheria,* but usually the main emphasis was on the raising of livestock that were supplied from the surplus at a mission.

The mission and the missionary performed many functions on the frontier. A mission increased the amount of land in the hands of a religious order, and after an area was no longer considered a frontier, the land often became quite valuable. It was against all rules and regulations for an order to own and sell land, but it probably did happen, and the funds would have been used to aid the mission system further out on the frontier. The priest was a civil administrator on the frontier. Most spent many years working among the Indians and spoke their language, and thus they almost always had the support and respect of the Indians. The Indians would trust the padre to settle their disputes and the disputes between them and the Spanish. The Spaniards saw no reason to separate church and state, and many religious leaders, such as archbishops or even missionaries on the frontier, held important and influential government positions. A missionary could also count upon his powerful friends, either in Mexico City or in Madrid, to right an injustice. He would only have to write an archbishop, or someone of authority in his society, in order to have pressure placed on the civil authorities. Thus the missionary wielded great influence on the frontier.

The missionaries also carried out many state functions. They explored and secured the frontier at little cost to the government. By controlling an area and by attracting Indians to the Spanish cause, they could prevent foreign encroachment, and the expansion into the New Mexico and California areas was primarily for that reason. The missions could also act as a buffer between hostile Indians and Spanish settlements, and they also promoted Spanish settlement because a mission provided a settled area into which settlers could migrate. Conflict between the Indians and the settlers was inevitable, and it was usually the missionary who was the arbitrator.

Missionaries added many things to the material culture of the Indians. The Indians had only a few domestic plants, but the missionaries introduced new field crops (especially wheat), garden crops (such as leeks, onion, cabbage, lettuce, garlic, carrots, and pepper), and new types of fruit (such as apple, quince, peach, apricot, plum, pomegranate, and fig). The missionaries taught the Indians how to plant, harvest, store, and cook the new crops. They introduced new agricultural implements and techniques, such as the plow, wagons and wheels, ax and adz, branding irons, and new irrigation techniques, and they also introduced new domestic animals. Prior to the coming of the Spanish, the only domestic animal most Indians of the Southwest had was the dog, which was generally not eaten, and some may have had domestic turkeys. The missionaries quickly and successfully introduced cattle, sheep, goats, horses, donkeys, mules, and chickens. The use of animal manure as a fertilizer for the fields was also introduced.

The missionaries also affected the Indians in a nonmaterial way. Teaching Christianity was their major goal, and once that objective was achieved, the Indian was greatly changed, since Christianity gave the Indians a completely new perspective on the supernatural and the afterlife. But the missionaries taught the Indians much more than religion. They taught the Spanish language and the rudiments of government, and they also taught many vocational skills, such as making adobe bricks and blacksmithing, that were valuable and needed around a mission.

Financing for the missions came from three sources: (1) state funding, (2) help from established missions, and (3) private sources. The first type of financing was very important. Because the government felt that it was in its interest to protect the mission and that it had a moral obligation to do so, it often supplied military support, and it was not unusual to find soldiers at a mission. The government also gave a stipend for each missionary in the field. This money was not given to the missionary but to the order or society to which he belonged, and that governing body, in turn, would disburse the money where it was most needed. The state also gave some money to each mission upon its founding. That money was used to get the mission established – to purchase such things as a bell, vestments, tools, nails, and domestic animals.

The state, of course, expected some things in return. The missionaries were expected to work with the Indians beyond the established frontier or, as the Spanish expressed it, beyond "the rim of Christendom." They were expected to teach the Indians Christianity and a knowledge of Spanish language and government, in order to help ease the transition when the Spanish settlers began moving in and the area came within the rim of Christendom. The Spanish authorities also expected loyalty to the crown. No one could accuse the missionaries of being simply tools of the Spanish government with the sole aim of advancing the cause of Spain. All of those men took hard vows and left an easier life in Europe to work for the rest of their lives in filth and poverty among the Indians. Many of the missionaries were not even Spaniards, as the names of some of those working in the Arizona area show: Keller, Grashoffer, Mittendorf, Sedelmayer, Pfefferkon, and Kino. As long as they were allowed to work with the Indians, the missionaries cooperated with Spain. The Spanish authorities, however, seldom completely trusted the missionaries or the societies to which they belonged.

The Spanish government also expected and required that when an area was settled and within the rim of Christendom (which theoretically took 10 years, but in the Arizona area took well over 100), the missionaries would leave the missions as established churches and move further out to begin their work all over again and the church and its land would be turned over to the secular clergy. Secular clergy are those priests who are ordained for a diocese and who work for a local bishop. Those priests who are trained and ordained by a society, such as the Jesuits, or an order, such as the Franciscans, do not labor for local bishops but rather for the order or society. (Orders and societies are very similar, and in this discussion they will be considered as essentially the same.) Priests belonging to orders take more vows than secular priests, for instance, they may vow poverty and obedience to the order, and they can be sent by their superiors to wherever they think the priests will do the most good—and that could be anywhere in the world. Since the mission buildings were to be handed over to the dioceses to become parish churches, the orders were not to benefit materially and the missionaries were to be forever missionaries. In the secularization of missions, the land and animals were also given up and divided among the converts, who then became citizens and landholders.

The new missions also received help from local missions that were already established. A successful mission would have a surplus of food, animals, clothing, and perhaps money that it would share with new missions. Even Indians who had been trained in technical skills, such as making adobe bricks, weaving, tanning, iron working, or carpentry, would be sent to help new missions get established. The mission system was meant to be self-perpetuating.

Financing was also received from private sources. Missionaries were constantly short of money for their work, and they had to rely heavily on private donations. This was known as the "pious fund" (fondo piadoso). The missionaries thus wrote lengthy letters to wealthy patrons, detailing the progress of the work and requesting contributions. It was a time-consuming but a necessary chore.

Spain used the mission system as its frontier agent to act as a buffer between the hostile Indians and the Europeans, and the missions prepared the natives for the day when they were to be a part of Spanish society with Spanish settlers living among them. Criticism can be made of the system, but it was a "humane" way of incorporating the Indians into the larger European society, and it was far different from the system practiced by the Anglos on the eastern seaboard, where "a good Indian was a dead Indian."

The greatest missionary on the Arizona frontier was Eusebio Kino. He was born in 1645 in the German-speaking Alto Adige, a region on the Italian-Austrian border now a part of Italy. During a critical illness, he prayed to St. Francis Xavier, his favorite saint, and vowed, if spared, to join the Society of Jesus. He survived, adopted Francisco as a middle name in gratitude, and joined the society in 1665. He was very well educated in German and Austrian universities and was knowledgeable not only in theology but in mathematics and astronomy as well.

Kino was determined to serve in the missionary field, and for eight years, he wrote letters requesting an assignment in China, the Philippines, or New Spain. He arrived in Mexico in 1681 and was assigned to work with the Indians in Baja California. Colonists settled La Paz, and Kino worked with the local Indians to try to help them make the transition to civilization. But the murdering of some Indians caused the settlement and the missionary to fail. Within a year, another settlement was attempted further north, but it, too, failed. Thus Kino's first missionary efforts were unsuccessful. In 1687, he accepted an assignment to work among the Pima, and this later work was to be very successful.

The area of his work was in present-day southern Arizona and northern Sonora, an area known as the *Pimeria Alta* that extended north from the Sonora River to the Gila River (Map 3.7). The western boundary was the Gulf of California and the Colorado River, and the easternmost boundary was the San Pedro River valley.

There were about 30,000 Indians living within this large area and they were divided into many groups speaking three basic languages (Map 3.8). The largest group were the Pima, who lived near the headwaters of the major streams of the area—the Altar, Sonora, San Miguel, San Pedro, Santa Cruz, and Magdalena. The Pima can be divided into the Opate Pima to the east, the Himeri Pima along the Magdalena, and the Pima Alto to the west. They were all basically agricultural Indians living in settled communities, and Kino was most successful with them. Another branch of the Pima Indians were the Sobaipuri, who lived along the San Pedro and the Santa Cruz rivers. They were more warlike than the Pima farther south, particularly those living along the San Pedro, probably because they were closer to the aggressive Apache. The Papago were a desert-dwelling people living in

PIMERIA ALTA

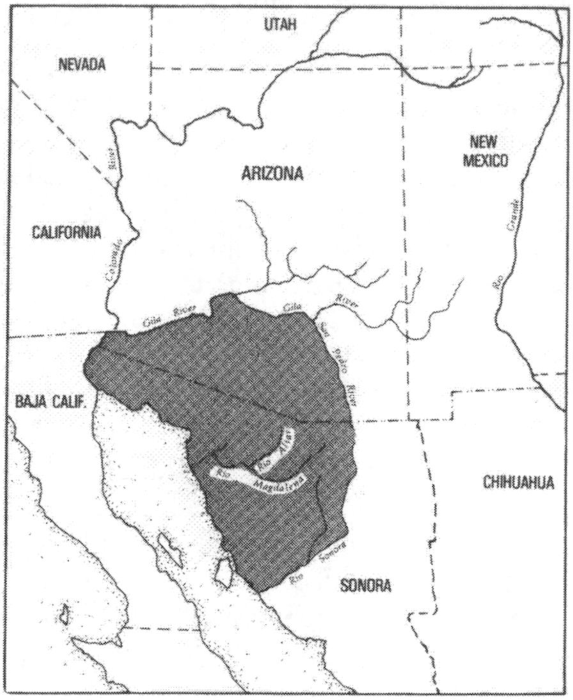

MAP 3.7

INDIANS OF THE PIMERIA ALTA

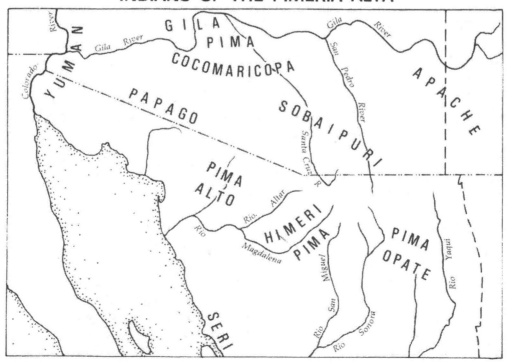

MAP 3.8

southwest Arizona. They were Uto-Aztecan speakers, like the Pima, but they practiced little agriculture and were basically hunters and gatherers with no permanent settlements. Therefore, Kino could not successfully work with them.

The second group, the Yuman speakers, lived in what is now southwestern Arizona. They included the Cocomaricopa who lived mixed with the Pima along the middle Gila, and the various Yuman speakers along the lower Colorado. They were agricultural Indians and were visited by Kino several times, but Kino never extended his mission system this far north and west. The Seri were also Yuman speakers who lived along the coast of Sonora, especially near Tiburon Island, but they were hunters and gatherers with whom Kino had no success. The last major Indian group occupying the Pimeria Alta were the Athabascan-speaking Apache, who were warlike hunters, gatherers, and raiders. Kino could never work with them, and they were to cause major problems for the entire mission system.

Kino's mission system can be divided into three groups: the core missions, the western missions, and the northern or Arizona missions (Map 3.9). The core group of missions was centered along the upper reaches of the Rio San Miguel, Rio Sonora, and Rio Magdalena. The mother mission was established in 1687 along the Rio San Miguel and was called Nuestra Señora de

MISSIONS OF PIMERIA ALTA

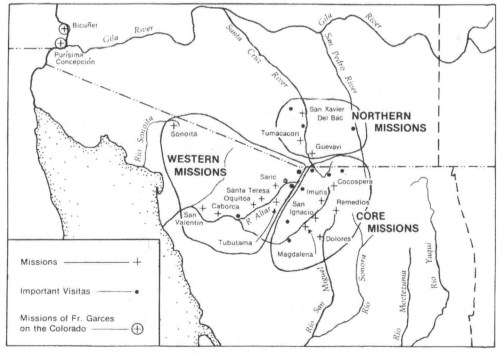

MAP 3.9

FIGURE 3.5. All that remains of the Dolores Mission.

los Dolores (Figure 3.5). Most of the missions had rather long names, but they were known by a shortened form; this one, for example, is usually called the Dolores Mission. It is built near a Pima village that had a headman of considerable influence, and this factor was a big help. The site was an excellent one. The Rio San Miguel was greatly constricted at the site, for the river had cut a narrow gorge through the rock (Figure 3.6), and just above and below the narrow canyon were extensive irrigable lands that were several miles long and about half a mile wide. The mission of Dolores was on the mesa on the west side of the river, and a sharp cliff faced the river and canyon. The church was thus in a high, commanding position, with good views of the farmland above and below, and it was easily approached only from the west. The mission's site and situation were excellent; there was plenty of easily irrigated land and an abundance of water, and the area was high and grassy, excellent for the grazing of livestock. The site could be easily defended, and it was inhabited by a large group of friendly, sedentary Indians. Within six years, the mission was obviously very successful. By that time there were irrigated fields, a mill, carpentry and blacksmith shops, much livestock, orchards, a vineyard, and a winery. Other missions were soon established, and all the core missions were successful.

The western missions were centered along the lower Magdalena River and its tributary, the Rio Altar. Probably the most important of these missions

FIGURE 3.6. The site of the Dolores Mission. At this point the San Miguel River, at the bottom of the photograph, is greatly restricted, and extensive areas above and below this point can be irrigated. Remains of the mission are to the left in the photograph.

was Nuestra Señora de la Concepcion del Caborca, or the Caborca Mission. The northern missions were in what in now southwestern Arizona (Map 3.9). The first mission established in that area was San Gabriel de Guevavi in 1692. For much of its early history, it was a mission in name only, and the first priest assigned to Guevavi did not arrive until 1701. He did not stay long because of ill health, and the next priest, Father Baptista Grashoffer, did not arrive until 1731. With a permanent missionary assigned to Guevavi, Tumacacori was made a *visita* in 1732. Kino visited Tumacacori many times, and while there he held religious services and encouraged ranching and farming. Tumacacori became a mission in 1752, long after Kino's death, and by 1784 it had become the major mission and Guevavi had been abandoned.

When speaking of the missions in Arizona, the first to come to mind is San Xavier del Bac (Figure 3.7). Kino first visited the site about 1692, and within a few years, the mission had become prosperous with many livestock, rich agriculture, and several thousand Indians. In 1700 a large church was started, but the first priest to remain over a year did not arrive until 1740. This mission grew to be very important, and many Indians moved to its site, particularly the Sobaipuri of the San Pedro who were very exposed to Apache raids. Several *visitas* were also established in the Arizona area, in-

FIGURE 3.7. San Xavier as it appears today.

cluding one along the San Pedro.

Although San Xavier was the northernmost mission, Kino traveled and preached to Indians far north of that point. He made at least six trips to the Gila River and two to the Colorado. Other Jesuits, such as Keller and Sedelmayer, also extended their work as far as the Gila, and Keller was given royal permission to work with the Hopi, since it was obvious that efforts by the Franciscans in New Mexico had failed. But Jesuit efforts to work with the Hopi also failed. They could not reach the Hopi because of the Apache, and mission efforts probably would not have succeeded anyway because of distance and the conservative Hopi attitude.

Kino was a man of great vision, industry, and drive, and he came to know the Pimeria Alta well. He made many exploring trips into present-day Arizona, sometimes with a military escort but often with only Indian companions. Most maps of his time showed Baja California an island, but he proved otherwise and drew a map showing that it was a peninsula. He was a good cartographer considering the tools he had to work with, and his maps of southwestern Arizona were widely published; for about 100 years, they were the primary sources for mapmakers (Map 3.10). Kino had plans to extend his mission system to the Gila River and from there to the lower Colorado. This would have provided a land connection with California, a major goal that was not to be accomplished in his lifetime.

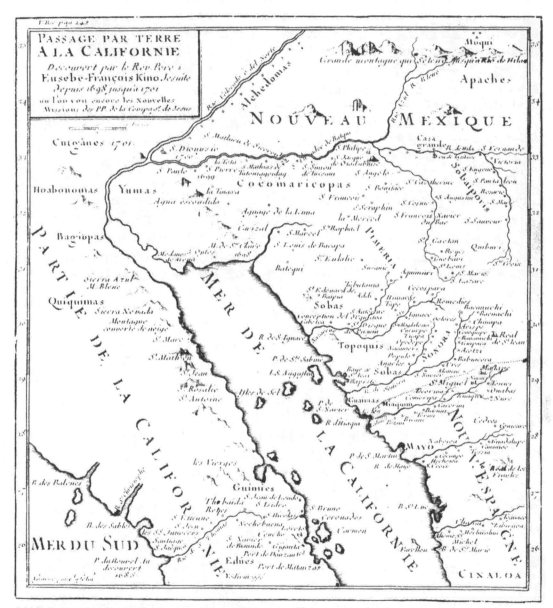

MAP 3.10. Map by Kino. Source: Ernest J. Burris, *Kino and the Cartography of Northwestern New Spain* (Tucson: Arizona Historical Society).

After 24 years of laboring in the Pimeria Alta, Kino died in 1711. He was buried at the mission at Magdalena, where his remains are now on display, and after his death the mission movement lost much of its vigor. A few new missions were established, but others were abandoned. One problem for the missions began in 1736 when an exceedingly rich silver deposit was found just south and west of present-day Nogales, in an area the Indians called *Arizonac.* Many miners and adventurers moved into the area, and they were a source of conflict. The strike, which consisted of large amounts of silver on or near the surface, was quickly exhausted, but unfortunately it revived the old, lingering rumors of gold and other riches on the frontier.

A severe blow to the mission system occurred in 1751 with the Pima rebellion that began along the upper Altar River but soon spread to other areas. A vast amount of damage was done to mission buildings, and some missionaries were killed. Military and mission authorities accused each other of provoking the revolt, and the exact cause has never been determined, but the Jesuits never recovered from the blow. The government, however, realized that it should establish a fort to represent its authority and to protect the missionaries and settlers. The military were not only to "show the flag" among the local Indians, but they were also to contend with the Apache, for by the middle of the eighteenth century, those Indians had become a real threat to security in the Pimeria Alta. A presidio was established at Tubac in 1752, three miles north of Tumacacori. The presidio was moved to an Indian village at Tucson for strategic purposes in 1776, and the villa of Tucson dates back to that time.

The end to the Jesuit missionary period occurred in 1767 when the Jesuits were expelled from all areas controlled by Spain – they had already been suppressed in France and Portugal before they were suppressed in Spain. Many charges were leveled against the Jesuits, but basically, they were considered too powerful politically and were thought to have conspired against the king. There was probably some truth to the rumors, because since the Jesuits drew much of their membership from the educated, wealthy, and politically powerful of Europe, they had many connections and were outspoken in their criticisms. All Jesuits in Mexico were taken into custody and put on ships for Europe, and the early Jesuit period in the New World ended. Many rumors of buried Jesuit gold began at that time. Supposedly, the Jesuits buried their gold, such as chalices and gold from some lost Jesuit mine, with the intention of returning for it. Unfortunately, Mexicans to this day have trouble with Americans digging up mission grounds to look for treasure.

The Franciscans were next given the task of doing missionary work in the Pimeria Alta. When they arrived, about a year after the expulsion of the Jesuits, the mission system was in disarray. Missions had to be rebuilt, and those missions that survive today were built during the Franciscan era. One of the Franciscans, Father Francisco Thomas Garces, followed in the tradition established by Kino.

Garces, like Kino, had large plans and traveled extensively on expeditions. He, too, wanted to open a road to California. By this time, missions had been established in California, and the authorities were receptive to the idea of opening a land route to that area. Garces, Capt. Juan Bautista de Anza, the commander at Tubac, and 34 soldiers made an overland trip to the California coast in 1774 and proved the practicality of the route, which developed into an important link between California and Mexico City. The only problem with the route was the crossing of the Colorado River, but the Indians there seemed friendly and helpful.

The Spanish authorities in Mexico City decided to establish two settlements along the lower Colorado, and both were built on the California side (Map 3.9). Father Garces was assigned to this area, and two missions were built. Married soldiers with families were assigned to this area, so the settlements were to be a presidio and a villa as well as missions, and they lasted for about a year until the Yuma rebelled in 1781. Many Spaniards were killed, including Garces. The Spanish were weak at that time, and the rebellion ended their land connections with California, as well as their efforts to work with the Indians along the lower Colorado.

In the late 1780s, a period of peace with the Apache began. The Franciscans successfully began to rebuild the mission system, and the number of Spanish settlers in the area increased. It was a time of relative prosperity, but the end came with Mexican independence in 1821. The Mexican revolution was a strongly anticlerical mestizo uprising, and it soon brought to an end the strong missionary pressure placed on the unwilling Indians, and Indian-mestizo-white relations became more flexible and tolerant. Many of the missionaries were quickly expelled as foreigners, and the missions began to fall into disrepair. In 1831, the Apache again went to war, and the Mexicans were even less able to cope with the situation than the Spanish had been. Three years later, the Mexican government secularized the missions, and the mission era ended.

The three missions in southern Arizona had varied histories. Guevavi was abandoned early, and little of the old mission survived. Tumacacori was slowly abandoned, and the local Indians took what movable property they could to San Xavier. In 1908 the old mission became the Tumacacori National Monument, and a partial restoration of the church was accomplished. San Xavier continued to deteriorate, but in 1913 the Franciscans reacquired it, and since then the mission has continued to serve the Indians.

MEXICAN AMERICANS IN THE SOUTHWEST AT THE TIME OF AMERICAN CONTACT

The settlement of California came late and was related to outside pressure, not to the discovery of gold or the possibilities for practicing agriculture. The Spanish never discovered gold in California – if they had, the map of the United States might be quite different from what it is today – and the Spanish felt no need for more agricultural land. The reasons for California's settlement are related to events in the Philippines. Spanish ships on their return from those islands sailed along the California coast and were easy prey for any pirates and enemy ships that lay in wait. The Spaniards, therefore, decided to settle and control California, and they moved quickly.

The coastal area had been explored by at least the 1540s, though it was over 200 years later when they decided to occupy the land. Settlement began in San Diego in 1763 with the establishment of a presidio and a mission. The Spanish were fairly weak by this time and could not afford great finanical expenditure, so the mission system was encouraged. For the next 40 years, Spanish settlers moved into California, and the mission system extended from San Diego to San Francisco. With Mexican independence in 1821, the secularization of the missions meant the end of the mission movement in California as well. California was a frontier area of a nation wracked by political agitation and revolution, and the people there received little attention or direction from Mexico City. There were troubles and turmoil in California also, and they continued until California was acquired by the United States in 1848.

The total Mexican-American population of California in 1850, when the first U.S. census was conducted there and very soon after the area became American, was 6,678, with about 2,500 of those having been born in California. The Mexican Americans made up only 7.2 percent of the total population – they were already engulfed by non–Mexican Americans – and the Mexican-American population was strongly concentrated along the coast from San Diego to San Francisco.

There were many Mexican citizens living in present-day New Mexico when the United States acquired that territory, and at that time, they constituted the largest concentration of Mexican Americans in the entire United States. In the 1850 census, they numbered 56,223 people, 91.5 percent of the territory's population. It is amazing how little they had expanded outward since the initial Spanish occupation in 1598, or in the 160 years since Diego de Vargas had reconquered the area. Their numbers had increased dramatically, but they had remained concentrated in the upper basin of the Rio Grande, with just a little spillover to the eastern side of the Sangre de Cristo Mountains. They were surrounded by roving bands of hostile Indians, and that fact largely explains why they did not expand outward. Indeed, expansion outward did not occur until American settlers and especially the American military had entered the area. They remained as an island of Mexicans in a sea of Indians. They were a small enclave, isolated from all civilization. The nearest Mexican town was El Paso to the south, but the route between the two, through the Mesilla Valley, was the dangerous *jornado del muerto.*

In Arizona, the Spanish and Mexican penetration had extended northward only to Tucson. The Spanish had attempted settlements and missions on the lower Colorado, but that effort had ended with the Yuma rebellion of 1781. There was a lull in the Apache wars that lasted for roughly 50 years, from the late 1780s to 1831. During that period of peace, there was population growth and a relative amount of prosperity in the northern part of the Pimeria Alta; the mission system flourished, towns grew, a few land grants were acquired, and ranching operations began.

With resumption of the Apache depredations, chaos developed. The ranchers that had begun operations along the San Pedro and Santa Cruz rivers were intimidated, and all abandoned their homes and cattle and left. Depopulation began, not just in present-day Arizona but in all of Sonora. A Mexican census dated 1848 counted only 760 inhabitants in Tucson and 249 in Tubac, for a total of only 1,009 Mexicans in Arizona. The population continued to decline, and in 1852 it was estimated at being about only 450. There were, therefore, no Mexicans living north of the Gila River when that area became American after the Mexican-American War, and at the most only a few hundred were living south of the Gila, all at Tucson or Tubac, when that area was acquired by the Gadsden Purchase.

SPANISH ADAPTATION TO THE SOUTHWEST

The Spaniards felt very much at home in what is now the American Southwest. California had a Mediterranean climate along the coast, and the Spanish understood that type of climate. The rest of the Southwest was primarily semiarid, and the Spanish also felt at home in that environment. Much of the Spanish Meseta is a high, dry, and rather barren place, so this new territory was not much different and in some areas probably more pleasing. The Spanish also had had a great deal of experience in the Sahara. Therefore, they knew what a harsh desert looked like, and they were not repulsed by the desert and semidesert conditions of the Southwest.

It is only natural for people to resist all change and to try to maintain their old way of life even in a new area. The Spanish were no different when they came to the New World. They attempted to establish the same social and economic conditions that existed in the Old World, and they were largely successful. What changes and adjustments had to be made were accomplished in Mexico, and expansion into the Southwest provided no break with that experience. It was just a natural expansion northward, with no difference between the new area and northern Mexico.

Spain still had a very feudal society when it explored, conquered, and exploited the New World. A few of the people were wealthy, well educated, had titles, and controlled the land, and the majority were peasants. The Spanish attempted, and largely succeeded, in establishing the same social system in the New World, with the Spanish and mestizo as the land-rich aristocracy and the Indians as the peasants.

The Spanish also established the same economy that existed in Spain. They brought over and planted crops they were familiar with, such as wheat, rice, barley, and various tree crops. Indian crops were not abandoned, however, and both sides accepted the other's crops. The Spanish were also familiar with irrigation practices in the Old World, and they passed on many of those ideas to the natives of the New World. The conquerors built buildings that were largely Spanish or Moorish in design, and adobe bricks worked as well in the New World as they had in the old. The livestock industry, with which the Spanish were very familiar, was transferred directly to the New World. It did change greatly through time, and it was well developed when it finally expanded into the Arizona area.

Therefore, in the Spanish settlements in the Southwest there were few breaks with the past. The people easily transferred to this area many ideas from Spain, such as construction techniques, social organization, the plaza, crops, and the livestock industry. Very few things were different in the New World, but one difference was that there was town planning. Most of the cities in Spain simply grew with little or no planning, and the streets wandered about with little or no order. But the Spanish were familiar with grid-pattern towns in the Old World, and almost all of their towns in the New World were laid out on such a grid pattern. This orderly adjustment did not occur in the more humid southeast, like eastern Texas, Louisiana, or northern Florida, where the Spanish settlers did not fare as well in, what to them, was an alien environment.

THE EARLY AMERICAN PERIOD

The Fur Trappers

The fur trade began along the East Coast of North America, and it was dominated by the French, for they held the good transport routes into the interior. Even when Canada passed into English hands, the Frenchmen continued to dominate the trade, only then they were working for the Hudson's Bay Company. By the 1750s, French trappers were at the headwaters of the Missouri in the Rockies, and soon after that they began visiting Santa Fe. The Spanish were alarmed and began handing out prison terms to the intruders. That practice kept the trappers out of Spanish towns but not out of Spanish territory.

Since most of the trapping took place in what the Spanish considered to be their territory, they attempted to get into the business but had no success. Few Spaniards were ever trappers. Only one person with a Spanish surname was important in the fur trade, Manuel Lisa, but even he was born and reared in New Orleans and was involved in the St. Louis end of the trade. The Spanish also tried to get the Indians to sell them furs, but they never made money on that trade. Any furs the Spanish in New Mexico exported had to go to Mexico City, and from there to Cuba, before they arrived in Spain, and the Spanish had little knowledge of the care, grading, or preparation of furs for shipment through the tropics. Nor did they have the connections to sell them. Others, therefore, continued to dominate the fur industry.

In the early American period, the American Fur Company in St. Louis dominated the trade. Rendezvous were held annually from 1825 to 1839, and the first four were held in Mexican territory, but far north of Mexican control. A rendezvous was a gathering of trappers, white and Indian, and merchants who brought goods overland in wagons from the Missouri River across the Great Plains and into the Rocky Mountains. A site was selected a year in advance, the word was spread, and the meeting was usually held in early summer. One important aspect of the rendezvous was that for the first time wagons crossed the Great Plains.

The opening of Santa Fe to American merchants was very important for the beginning of the fur trade in Arizona. Spanish officials had always kept the Americans out, but with independence in 1821, Mexican officials began allowing the entrance of American merchants. There was soon a flood of goods moving between Santa Fe and the United States, all of it following the Santa Fe trail that originated in the Kansas City area. The Americans greatly undersold their Mexican competitors because Kansas City is much closer to Santa Fe than is Mexico City and many of the manufactured goods sold from Mexico City came from Europe. Also, American goods were often of much better quality. Fur trappers soon followed the merchants. The Mexican authorities required licenses of the trappers, but many did not bother with this formality.

The Arizona fur trade began comparatively late and developed slowly. Most of the fur trade was centered on beaver pelts, and Arizona was not particularly good beaver country (Figure 3.8). Also, its geographical position in relation to the main centers of the trade was not a good one. The Arizona fur industry, therefore, faced many problems. There were few streams, and many of those that did exist were too shallow or erratic in their flow to support beaver. It was also a long overland route to St. Louis, the major processing center, and it was difficult, if not impossible, to bypass Santa Fe, where the Mexican officials taxed and charged for licenses which meant there was less profit to the trapper. There was great variety in the quality of the beaver pelts that came from Arizona, ranging from good to inferior. Nevertheless, there was money to be made in the Arizona fur industry.

The first major fur trapping expedition into Arizona occurred in 1824, and it was led by a father-son combination, Sylvester and James Ohio Pattie. They trapped along the upper Gila River and its tributaries, particularly the San Francisco and the San Pedro. Their party had Indian trouble and had to cache their furs and return to New Mexico empty handed. When they returned to retrieve the furs, they found half of them destroyed, but a large profit was made on those that remained.

FIGURE 3.8. A beaver dam and lodge along the upper reaches of the Little Colorado River.

The height of the beaver trade in Arizona was reached in only two years. By 1825 Americans, many of them of French descent, were trapping and retrapping the streams in Arizona, and soon there were very few beaver left in the area. The industry had abandoned the area by 1831, and the last beaver expedition was sent to Arizona by the Hudson's Bay Company in 1844, and it was a failure.

The significance of the fur trade was that for the first time Americans were entering the area. They had a great impact on the Americans who arrived later, for they acted as guides to the military, settlers, explorers, and miners who were to come into the area. Three of the more-important and better-known trappers were Bill Williams, who has a river and a mountain named for him; Pauline Weaver, a half-Cherokee from Tennessee who led mining parties to important gold discoveries and who has a peak in the Superstition Mountains named for him; and Antoine Leroux, who worked in northern Arizona and was a guide for the Mormon Battalion. These men were very important in the opening of Arizona to the Americans.

American Acquisition of Arizona

Most of Arizona became American as a result of the Mexican-American War, but the area played no role in that war, and it was not by design that the United States acquired part of Arizona because of it. The origins of the conflict can be traced to a variety of factors, including the Texas western boundary question (Texans claimed the full length of the Rio Grande as their western border), the desire of Americans to acquire California, and the American ideas of "manifest destiny." Mexico claimed vast amounts of land but did not occupy or exploit it, and therefore the Americans felt that Mexico did not have a strong title to the land.

James K. Polk had strong ideas about manifest destiny, and he considered his election to the presidency in 1844 a mandate to carry them out. He made overtures to buy the land west of Texas, but the Mexicans refused to sell. In April of 1846, Mexican soldiers attacked American troops north of the Rio Grande, in territory claimed by both governments. That was the excuse for which Polk was waiting; he went before Congress, said Americans were attacked on American soil, and on May 13 war was declared. Mercifully, it was a short war, because it received very little support from the American populace and had it dragged on, it would have been difficult for the Americans to win.

There was no fighting in the Arizona area, which was considered just a big desert between the Americans and their goal, California. It had no cities, large populations, rich agricultural areas, or known mineral wealth, and its only strategic consideration lay in its position – it was in the way. The war was to be decided in Mexico, and any military failures or successes in the American Southwest would have little impact on the outcome. The Americans landed a force in Mexico, conquered Mexico City in September of 1847, and the war came to an end.

Two military expeditions came to the Arizona area, not to occupy it but only to travel through it. One of these expeditions was led by Col. Stephen Watts Kearny. He was authorized to organize the "Army of the West," which he did in western Missouri. It was a miscellaneous force made up of three main elements: 300 army men, 500 Mormons, and 1,000 Missouri adventurers. The regular army men were trained and dependable. The Mormons joined because at that time they were preparing for their migration westward from Illinois, and even though they felt animosity toward the American government, they saw this as an opportunity to get some of their men to the west and to use their pay to help finance the move.

Since the Missourians were not trained soldiers, there were questions as to how they would react in combat, but later, in Mexico, they proved to be very effective fighters. Tagging along with the army were several hundred traders and travelers going to Santa Fe. The Mexican governor at Santa Fe surrendered peacefully, and Colonel Kearny claimed the area, which at the time included most of Arizona, for the United States.

Kearny did not waste time, and he immediately began making plans for continuing to California, the real goal. The men heading for California were split into two groups; one group was to head directly to California and the others were to take a more southerly route with the supply wagons. Colonel Kearny led the 300 dragoons that were to head directly for California. They headed south and then westward toward the Gila River. While still in New Mexico, they met Kit Carson going east with dispatches stating that the Americans had already taken California. Kearny reassessed the situation, sent 200 of the dragoons back to Santa Fe, and with the remaining men and Carson as a guide, Kearny headed west toward California. The group followed the Gila River through Arizona (Map 3.11), but it proved to be a difficult and impractical route and was never again followed by travelers crossing the area. They found the Indians in Arizona to be friendly, even the Apache, who considered anyone fighting the Mexicans to be their ally. Kearny and his men continued on to California. They arrived near the coast in a debilitated condition from the long march, and this, along with confusion, led to their defeat in the Battle of San Pasqual, though they considered

ROUTES OF COL. KEARNY AND THE MORMON BATTALION

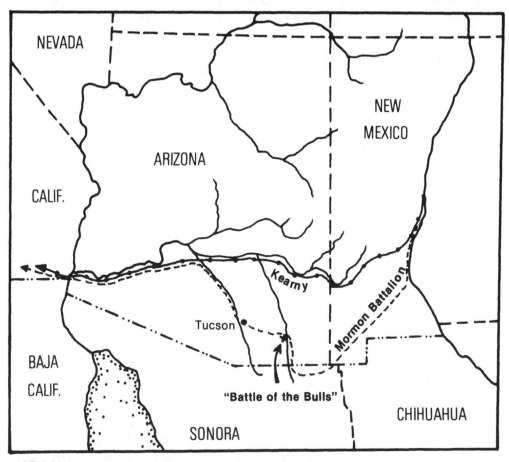

MAP 3.11

it a victory since the Mexicans retreated from the field.

The other expedition across Arizona was the Mormon Battalion led by Capt. Philip St. George Cooke. This army consisted of about 500 men plus 24 supply wagons. Their guides, former fur trappers Antoine Leroux and Pauline Weaver, realized that the Gila River route would be inappropriate for the wagons, so they left the Rio Grande and headed westward much farther south. Their route went through southern New Mexico, then south into Mexico, and then into Arizona along the San Pedro River (Map 3.11). It was along the San Pedro that the Mormon Battalion fought its only battle, known as the Battle of the Bulls. There were many wild cattle that had been abandoned when the Spanish ranchers were forced from the area, and it was some of those wild bulls that attacked the battalion. Several men were hurt and two of the mules were killed, but the "battle" was easily won. They reached Tucson in mid-December. The Mexican garrison there offered no resistance and retreated from the town. The Mormons rested in Tucson and then continued the journey, reaching San Diego well ahead of schedule. Cooke was not the least impressed with Arizona and had little good to say about it. The most significant aspect of this trip was that for the first time wagons crossed Arizona. It was clear from Kearny's experience that wagons could not follow the upper Gila, but a suitable route had been proved to exist to the south, and many wagons later used Cooke's Wagon Road, which became known as the Gila Trail. When that area was not made part of the United States after the Mexican-American War, many people were upset, and it was an important factor in the decision to acquire more area to the south.

The Treaty of Guadalupe Hidalgo, signed on February 2, 1848, officially ended the Mexican-American War. A vast portion of Mexico was ceded to the United States for $15 million, including almost all of Arizona north of the Gila River. In the treaty, the Americans agreed to help stop Apache depredations into Mexico, something the Americans did not accomplish for a long time. All conflict between the two nations had not ended, however, for the southern border had yet to be identified.

According to the Treaty of Guadalupe Hidalgo, the border between the United States and Mexico was to follow the Rio Grande to El Paso, then turn westward and finally due north until it intersected the first branch of the Gila River. It was then to follow the Gila to the Colorado River. The Gila River section of the border posed no problem, and it was surveyed, but the area between the Rio Grande and the Gila River caused problems and hard feelings. According to the treaty, the border was to be laid down according to the Disturnal Map. This map had been published in New York in 1847 and was titled "Map of the United Mexican States," but it was not accurate and was little more than a copy of an 1822 map. According to lines of longitude and latitude on the Disturnal Map, El Paso was 34 mi. (54.7 km.) too far north and 100 mi. (160.9 km.) too far east. It is possible that the Mex-

icans knew this when they had the Disturnal Map included in the Treaty of
Guadalupe Hidalgo, for the Mexican representative to the border commission, Gen. Pedro Garcia Condé, insisted that the boundary begin according
to where El Paso was on the Disturnal Map, not where El Paso actually was.
The American counterpart to Condé, John Russell Bartlett, found Condé
adamant, but finally a compromise was agreed upon. The line would begin
according to the Disturnal Map but would turn north 3° west of where the
Rio Grande was in reality (Map 3.12). When news of the compromise
reached Washington, D.C., the Whigs were in power, and it was perfectly
acceptable to them, but the Democrats, especially the Texans and Southerners, were very upset. They saw the Whigs giving away 6,000 sq. mi.
(15,540 sq. km.) of land that should belong to the United States. First, that
land included much of the Mesilla Valley (just north of El Paso on the Rio
Grande), a potentially rich agricultural area that was as yet unoccupied, and
second, it included the only proved wagon and rail route between the South
and the West Coast. Through political maneuvering, the Whigs managed to
block the surveying of the Bartlett-Condé line between the Rio Grande and
the Gila River. The governor of Chihuahua considered the disputed area
part of Mexico, and troops were sent in to briefly occupy the area; the governor of New Mexico sent a force to the Mesilla Valley and claimed it for the

SOUTHERN BOUNDARY

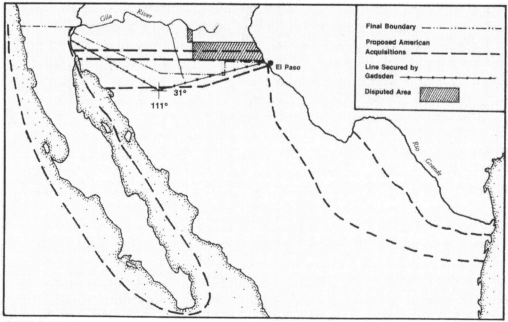

MAP 3.12

United States. A second Mexican-American War seemed in the making.

The Democrats came back into power when Franklin Pierce was elected president in 1852. One of the strongest advocates pushing for more land in the Southwest was the Southerner and Secretary of War Jefferson Davis who was particularly interested in a good rail route between the West Coast and the South. Both countries, however, wanted a peaceful solution to the border problem; another war would have badly split the United States, and the Mexican dictator, Antonio Lopez de Santa Anna, desperately needed money. James Gadsden of South Carolina was sent to Mexico to buy land to settle the dispute between the two countries. He had orders to get as much land as possible—for example, northern Chihuahua and Baja California—but at all cost he was to get more land between the Rio Grande and the Colorado River (Map 3.12). The Mexicans would not part with Baja California, and they were determined to prevent the Americans from acquiring a port on the Gulf of California, which would have cut any Mexican land connection to Baja California.

Gadsden struck a bargain with the Mexicans in the Treaty of Boundary and Cession of Territory, which was signed on December 30, 1853. President Pierce was pleased, but the Senate was not ready to agree to the terms of the treaty. This was just prior to the Civil War, and the Senate saw no reason to buy more land just to please the Southerners and add land and power to the slave states. After much bickering, the treaty was finally passed in an amended form, and the Americans got most of the land for two-thirds the cost, $10 million. The boundary agreed upon gave the Americans control of a good transcontinental route. The border began just above El Paso, and so included the Mesilla Valley, and went westward 100 mi. (160.9 km.). It then jogged south for 20 mi. (32.2 km.) to include several important passes and roads, especially Guadalupe Pass in the southwest corner of New Mexico, and then, just beyond the headwaters of the Santa Cruz River, it angled toward the Colorado River.

Early American Development

New Mexico Territory at first included what is now part of Arizona, but with the land added by the Gadsden Purchase, the geopolitical character of the area began to change. The new land had traditionally been governed from Sonora to the south, not from Santa Fe, and the Mexican settlers who had moved north did not want to be controlled from Santa Fe. Many Americans began moving in, particularly Texans, and any expression of power emanating from Santa Fe was greatly resented by the Texans who never did trust the Hispanics in Santa Fe. Agitation for status as a separate territory began immediately upon ratification of the treaty.

The first efforts to make Arizona a territory would have given Arizona an east-west orientation (Map 3.13). A petition from Tucson in 1854 requested that a new unit be created out of southern New Mexico and be given the name "Gadsonia," "Pimeria," or "Arizona." Similar requests also came from the Mesilla Valley, and there was also congressional support for such plans. A Tucson convention in 1860 set forth territorial bounds, separating Arizona from New Mexico along latitude 33°40′N, and that boundary seemed a strong indication of things to come. When the Union split and the Confederates had a force in the Southwest, such an Arizona territory – with its boundary a bit farther north, along 34°N latitude – was formally recognized by the Confederate States of America. An east-west Arizona territory was not out of touch with reality. Santa Fe had exercised some control in northern Arizona, especially over the Hopi area, but had never extended that control to southern Arizona, nor had Santa Fe extended its control to southern New Mexico in the area of the *jornado del muerto*. Also, southern New Mexico was largely settled by Texans who did not want to be governed by the hispanics in Santa Fe. The western boundary of this confederate territory ran through the rugged and largely unexplored, unmapped, and unexploited White Mountains and the Mountain District just south of the Mogollon Rim. These boundaries provided a southern corridor and thus fit well with Southern strategy. Washington, understandably, would not accept these ideas and divided the large New Mexico Territory in a north-south fashion along the 109°2′59″W meridian – in direct opposition to the Southern plan. Thus, Arizona was formally attached

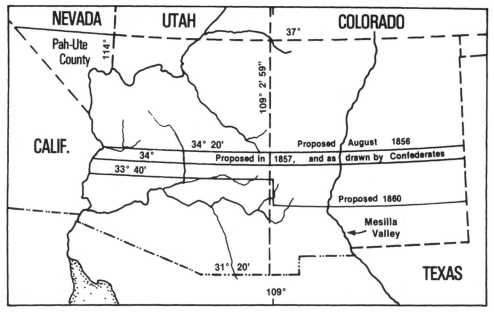

PROPOSED ARIZONA BORDERS

MAP 3.13

to the Union and became a territory in 1863.

The northern boundary was established along the 37th parallel. Because of its predilections for symmetry, Congress drew a straight line, never considering the physical or cultural factors, and troubles developed because of that line. At the time the territory was organized, the northern border extended all the way to the California line, creating a sharp-angled northwest corner for the territory. Legislation establishing Nevada provided for potential expansion eastward and westward. It could not expand westward because of the political clout of California, but the border was shifted eastward at the expense of Utah. In 1866, Congress shifted the Nevada boundary eastward again, but it also extended the Nevada line south to the Colorado River. Utah and Arizona complained loudly, but they were defeated by the mining interests in Nevada. Because of the shift, Arizona lost Pah-Ute County, which had been established in 1865 (Map 3.13). Arizona territorial legislation in 1871 abolished Pah-Ute County but did not relinquish the land, for it stipulated that any territory that might remain in Arizona be a part of Mohave County.

Utah, meanwhile, had designs on northern Arizona and tried on several separate occasions to acquire the Arizona Strip. In 1865, the Utah territorial legislature petitioned Congress for the Strip. Mormon settlers had been expanding southward into the Virgin River Valley, and a steamboat landing was planned at the junction of that river and the Colorado at Callville. An important part of the Utah petition for the Strip was the fact that Utah wanted to have a port on navigable waters, and Callville at the time was still in Arizona. Congress did not act, and the matter died. In 1902, after Utah had been made a state and Arizona was still a territory, Utah again petitioned for the land with no result. Utah then tried another tactic to acquire its goal. In 1903, a commission was sent to Arizona to request consent for the Utah annexation, but quite expectedly the request was firmly rejected. Utah tried again in 1909, but for the third time, congressional inaction left the area as a part of Arizona. Utah made one final attempt to acquire the Strip in 1919. At that time, Utah was prepared to buy the land or trade it for Utah land south of the San Juan River along the northeast border of Arizona, but again Arizona rejected the idea.

Utah made some strong agruments for acquiring the Strip, and if it had been made a part of Utah, the Colorado River would have made a good cultural and physical boundary. Some of the arguments for Utah's acquisition were (1) almost all of the inhabitants of the Strip were Mormons with close ties to Utah, (2) it was mainly a grazing area used by Utah residents, and (3) it was a no-man's-land as far as justice was concerned because Utah had no jurisdiction there and Arizona peace officers were mainly on the other side of the Grand Canyon. But Arizona was adamant about keeping the Strip and managed to do so, but the cultural and economic situations have not changed.

After American acquisition came a period of American exploration. The entire area had been explored by the Spanish in a cursory way, but the new owners were interested in learning more and in mapping the area. American explorers only retraced the routes of the earlier explorers, and they really discovered little that was new. The only area of Arizona left to be truly "discovered" was the bottom of the Grand Canyon, and the exploration of that area was led by a one-armed Civil War veteran, John Wesley Powell, who began his exploration in 1869. The federal government was interested in tying California to the eastern part of the country, and the location of routes westward was considered critical. The Gila Trail followed by Cooke's wagons was well known, and thousands had used that route in going to the California gold fields in 1849, but no other routes through Arizona were well known. As a result, almost all early exploration was for the purpose of blazing routes westward and was conducted by the United States Army Corps of Topographical Engineers.

Several military expeditions crossed Arizona prior to the Civil War, and most were on reconnoitering expeditions to search for suitable routes westward. The route followed today by the Santa Fe Railroad was explored by Capt. Lorenzo Sitgreaves and later by lieutenants Amiel Weeks Whipple and Joseph Christmas Ives. A later survey along this route, led by Naval Lt. Edward Fitzgerald Beale, involved the famous and unique Camel Military Corps. On that trip, camels proved their effectiveness in the Southwest, but they were too different from the animals with which Americans were familiar, and their use was eventually rejected. There was also one military expedition across southern Arizona, led by lieutenants John G. Parke and George Stoneman, to survey a route now essentially followed by the Southern Pacific Railroad. In this way much of Arizona was mapped and became known.

There were many proposals for the construction of a railroad across the western desert in the 1850s. A southern railroad route through Arizona received a great deal of support, and it would have had several advantages over a more northerly route: (1) it was the shortest route between the East and the Pacific, (2) it would have been considerably cheaper to build than other routes, (3) the winters were mild so there would have been little danger of snow closing the route, and (4) there were coal fields along the Brazos. The major drawback, however, was it would have been a very circuitous route to San Francisco, the major destination, and most heading for California were from the Midwest and desired a more direct route. When the first railroad was finally built, it was just after the Civil War, and the feeling then was that the railroad had to originate in a "Northern" city—even St. Louis was too "Southern." Therefore the first railroad west, the Union Pacific, began at Council Bluffs, Iowa, which had good connections due eastward to Chicago, and from there it headed west to San Francisco.

Two early routes were prosposed across Arizona, one along the 32nd parallel and the other along the 35th parallel. These routes were not meant to serve the territory, only to cross it. The Southern Pacific, which tended to follow the 32nd parallel, had the earlier start; it began building eastward at Yuma in late 1878. This railroad followed the one well-known, traveled route through the state, the Gila Trail, and deviated from it only in southeastern Arizona (Map 3.14). Most east-west travelers, and the route of the Butterfield Stage, went through Apache Pass between the Chiricahua and Dos Cabezas mountains. Since that time, all traffic, such as Interstate 10, has followed the Southern Pacific route, and Apache Pass is now an isolated and seldom-visited spot.

The 35th parallel was followed by the Atchison, Topeka and Santa Fe Railroad. This railroad ran from eastern Kansas to New Mexico, but once in New Mexico, promoters of the Santa Fe had grander schemes and desired access to the Pacific Coast, so the Santa Fe extended a line to southern New Mexico where it linked with the Southern Pacific line. By using Southern Pacific lines and building southward in southeastern Arizona, the Santa Fe achieved its goal of reaching a western port, but the port was Guaymas, in Mexico and on the Gulf of California. This move was largely a result of being blocked from expansion due west, but it did result in a very short transcontinental route. It also, perhaps, reflected some of the American attitudes of the time that northwestern Mexico was simply a part of the greater Southwest, due to be tied to it economically if not politically.

In 1880, the Santa Fe gained control of an old Atlantic and Pacific Railroad project—that company had acquired the rights to build across Arizona—and immediately began building westward along the 35th parallel. In doing this, the Atlantic and Pacific retained title to over 11.5 million acres (4.657 million ha.) of land between Albuquerque and the California border, land that the federal government had given as an inducement to build the line. Railroading in California was still monopolized by the Southern Pacific, however, so once its tracks had extended to the Colorado River, the Santa Fe was again blocked from further expansion westward. An agreement between the two railroads was finally reached, and the Santa Fe was given a route to the California coast in exchange for the Guaymas route (Map 3.14). The Santa Fe got far and away the better part of that deal, for it received access to California and the Southern Pacific got a relatively unimportant line, which it lost when Mexico nationalized its railroads. Thus, by 1885 two transcontinental rail lines extended across Arizona. Another rail line of some consequence to Arizona was a spur built by the Santa Fe in the 1890s that connected its main line to Phoenix where it linked with a Southern Pacific branch line that extended northward from Maricopa. This linkup provided the only north-south rail line in Arizona. The last significant rail link to be built was the Parker Cut-off, a Santa Fe line that extended from Phoenix through Parker and gave the Santa Fe excellent connections between Phoenix and Los Angeles. The remainder of railroad history in Arizona is mostly one of building small feeder lines to mines and outlying towns.

Although these two railroads were built only to cross Arizona, their

IMPORTANT EARLY NON-MINING RAILROADS

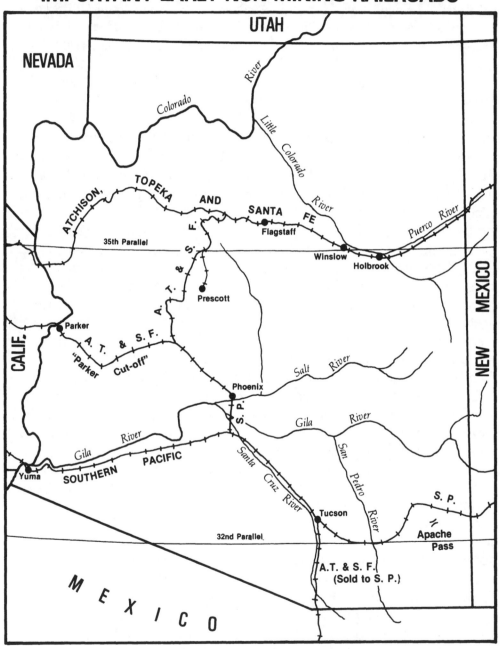

MAP 3.14

economic impact on the territory was tremendous. Prior to the railroads, the only cheap mode of transport in the territory was steamboat travel on the Colorado River (Figure 3.9), but that area was far from the center of economic activity. It was, therefore, the railroad that really opened the territory to development, since it allowed numerous travelers, settlers, and merchants to enter Arizona quickly, easily, cheaply, and safely. There was an immediate impact upon those industries that produced large-bulk but low-value products, such as the lumber, livestock, and mining industries. The mining industry was especially affected, for the railroads entered the territory during the time when high-grade silver and gold ores were being depleted and copper, a cheap ore requiring inexpensive transport, was gaining importance. It was not long before every mining district had rail connections to the transcontinental routes.

The military also played a major role in the early settlement of Arizona. The control and development of Arizona were major goals of the American government, and it was left to the military to provide the leadership needed to solve many of the early problems. Once settlements had been built and settlers had begun to exploit the area, the military provided protection from the hostile Indians. Another mission of the military was to prevent Apache

FIGURE 3.9. An old steamboat on the Colorado River. According to early treaties, Mexico gave the United States free access to the Colorado River, and during the steamboat era many towns were established along the river, though few such towns still survive. Courtesy, U.S. Bureau of Reclamation.

raids into Mexico. This they were required to do according to Article II of the Treaty of Guadalupe Hidalgo, but it was to be many years before they succeeded in that task.

Much of Arizona's early economy was tied to the military, which in the 1850s and 1860s made up a considerable portion of the population. For example, in a special 1864 census, 23 percent of all people enumerated were military personnel. For many years, the money that was spent to maintain and support this army was the major economic mainstay of Arizona. Army payrolls helped to support many towns, and citizens protested over the closing of any military base. The citizens of Tucson, for example, complained loud and long when Fort Lowell was abandoned. Many merchants made money hauling and selling goods to the military, and Phoenix originally came into existence as a farming community to supply Fort McDowell with hay. Arizona thus depended greatly on the military, and when the Indian wars ended in the 1880s and the military began to withdraw, Arizonans sorely missed the army payrolls and the contracts to supply the military with food and merchandise.

The military also made life better for the civilians in other ways. Well-educated officers and their socially conscious wives contributed much to the social life in isolated frontier towns, since they gave parties and organized events such as concerts by the military bands. In the area of medicine the military also contributed significantly, because most of the early doctors were military men and most of the hospitals were run by the military.

The military personnel were a varied lot. The officers were mostly well-educated West Pointers who desired a tour of duty in the West, because that was the only place they could see military action and thus further their careers. The enlisted men were just the opposite of the officers. Most were uneducated and foreign-born (many in Ireland or Germany), and they were in Arizona only because they had been sent there. Black troopers (called buffalo soldiers) served in Arizona in the 9th and 10th Cavalry and 24th and 25th Infantry, which were led by white officers. These black soldiers performed outstanding service in the field and had an extremely low desertion rate. Overall they were probably the best soldiers in the Southwest.

Americans have always had a love-hate relationship with the military, and that was certainly true in the early history of Arizona. Whenever there were Indian problems, the civilians immediately demanded military support. The civilians seemed to care little for the fact that the enlisted men had a hard and an unrewarding life – living in crude shelters and often on long and dangerous patrols – or for the fact that life at an isolated post in Arizona must have been extremely monotonous. There were also social prejudices against the enlisted men, and the blacks especially were ostracized. The officers were better received socially, but many commanders were heros one day and lambasted by the press or some politician out for votes the next. The situation is the same today: the civilians complain about the military, but the threat of losing a local base will have most nearby civilians opposed to the closing.

The military were sent to northern Arizona after the Treaty of Guadalupe Hidalgo and to southern Arizona after the Gadsden Purchase. There were, however, only a few problems with hostile Indians in the 1850s. The Navajo resisted white settlers and travelers, and Fort Defiance, the first American military post on Arizona soil, was built in 1851 to counter that threat. The Yavapai and some of the other Indians were involved in the killing of some whites, but the Apache generally left the Americans alone and raided deep into Mexico instead. But the situation changed in 1861 as a result of the "Bascom affair," in which several Indians and whites were killed. Cochise, the leader of an Apache group in southeastern Arizona, then began a bloody war that was to take hundreds of lives in an effort to drive the whites out of southern Arizona. Unfortunately for the Americans, the Civil War began before the army could begin an offensive operation against the Apache, and the troops were needed in the East. Many forts were abandoned and burned and much blood was shed during the 1860s, and the Apache thought they were winning. The Navajo were subdued by Kit Carson in 1863–1864, but the Apache continued to resist long after the Civil War. However, the American military began to assert its authority in the 1870s, and by the mid-1880s, the last of the Indian revolts were over.

The military forts in Arizona and elsewhere in the West were very different from what is imagined by the average American. Despite what the movies or television serials might portray, they were not compact forts with high walls around them to keep out the Indians and portholes from which to fire rifles. On the contrary, they were open and expansive (Figure 3.10). This type of fort was first built along the edge of the Great Plains and then further west. A few forts deep in Indian territory did take on the appearance of a walled fort, for as more buildings were added, they were squeezed in between existing ones and they appeared to have a common wall facing outward, but their basic design was still that of the open fort. Even though the forts were quite open, they were very seldom bothered by the Indians. Hostile Indians were organized into small bands of related individuals, and no Indian wanted to be hurt or killed or have that happen to a relative, so attacks were long deliberated and were usually little more than hit-and-run raids. The Indians just did not attack forts full of well-armed and disciplined men; instead, they usually attacked isolated farmsteads or travelers. There is an exception to every rule, and in 1860 over 1,000 Navajo did attack Fort Defiance—and were severely thrashed.

The average military fort was centered around a large parade ground, which was either square or oblong and had a flagpole in the center or to one side. A road ringed the parade ground, and a road or two, plus trails, led away from the fort. The officers' quarters were on one side of the parade ground. Many of the officers had their wives and children with them, and the officers' quarters were usually quite nice, single-family residences generally resembling some of the finer homes found in the midwestern or eastern cities at that time. This section of the post usually had some trees,

A TYPICAL FORT IN EARLY ARIZONA

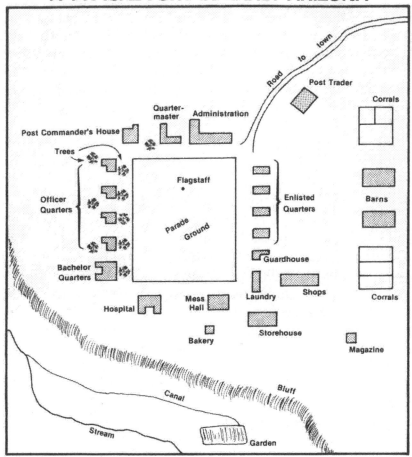

FIGURE 3.10

because the officers' wives would have them planted soon after their arrival. Usually the barracks of the enlisted men were across the parade ground from the officers' quarters. On a third side of the parade ground were the buildings associated with the running of the fort, such as the commander's office and the quartermaster's office (Figure 3.11). The fourth side also usually had such buildings, like the mess hall, bakery, commissary, recreation room, laundry, guardhouse, and so forth. The horses were important, but because of the flies and smell that they generated, plus the fact that it was the enlisted men who attended to the horses, the barns and corrals were away from the parade ground and nearer to the enlisted quarters than to the officers' quarters. The buildings of most of the forts in the southern part of Arizona were of adobe, and those in the north

FIGURE 3.11. Former adjutant's office at Fort Apache. It is made of adobe.

were of adobe, brick, or wood.

Most of the military posts of Arizona were located in or near the Mountain District and in the Mexican Highlands (Map 3.15) and were thus concentrated in a band from southeast to northwest. The exceptions were four posts along the lower Colorado and four posts in the northeastern section, which were established to control the Navajo. It must be remembered that these forts did not all exist at the same time but were built, used, and abandoned during various stages of Arizona's history. Some of the military establishments were designated as "camps" instead of forts, and others were known by such terms as "post," "cantonment," "barracks," and "presidio." Many designations were changed from camp to fort and vice versa, and many posts also changed names over the years, as well as locations.

The prime consideration for the location of any fort was its role in controlling the Indians. Many forts were located along streams in order to deny the hostile Indians access to water, and others were built along important passes or astride routes leading into Mexico to prevent Apache raids south of the border. Still others were built in the center of Indian country, such as Fort Apache, which was 30 mi. (48.3 km.) from the nearest town and had no road to it, and some were established near white settlers to afford them protection. Once the general location of a fort had been decided upon, a local

IMPORTANT EARLY MILITARY POSTS IN ARIZONA

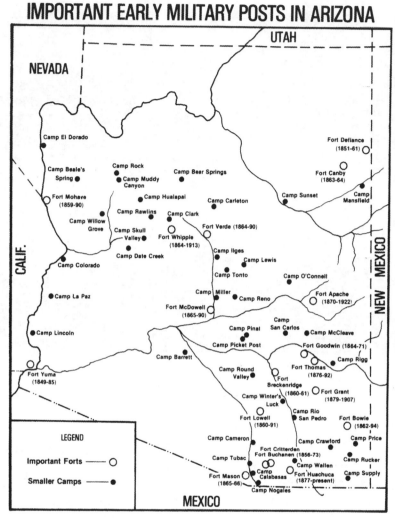

MAP 3.15

commander selected the actual site for the post. Many factors were considered in this site selection, and an especially critical factor was access to dependable water. However, if the post was to be in a valley, relatively high land would be chosen so there would be fewer problems with disease or flooding.

No significant military posts were established after 1877, and after 1886 and the collapse of the Apache threat, most of the forts were rapidly disbanded. Many of the forts were auctioned off, some were simply abandoned, and ironically, others were given to the Indians they had been built to subjugate and are still in their hands. The only fort to maintain its military function into the twentieth century is Fort Huachuca. It was a focal point of activity during World War I and was a training base for black soldiers during World War II.

4

POPULATION AND ETHNIC GROUPS

GENERAL ASPECTS OF THE POPULATION

The population of Arizona is heterogeneous (Figure 4.1), though whites compose 72 percent of the population. The second largest segment of the population is made up of people of Spanish heritage. That group constitutes 18.7 percent of the population, but some argue that this figure is too low, for many persons in this group are not counted, particularly the illegal aliens who made an understandably determined effort to avoid enumeration. The third largest group, the Indians, composes 5.6 percent of the population, and the blacks, 3 percent. The remainder of the population, classified as "other," is figured at 0.7 percent of the overall population.

The population of Arizona has grown rapidly in recent years. Between 1970 and 1975, it increased at a rate of 25.3 percent, while the national growth rate was only 4.1 percent. This rate of growth was the highest in the nation, and the only other states to approach the growth of Arizona were other Sun Belt states, such as Florida with 23 percent, and Nevada (with a much smaller population so the growth rate reflects considerably fewer people) with a growth rate of 21 percent.

Rapid population growth is not a new trend in Arizona, for it has experienced rapid growth ever since it became a part of the United States (Figure 4.2). The first census conducted in Arizona, soon after it achieved territorial status, counted 4,573 persons. By 1870 the population had increased to 9,658, and between that time and the 1880 census, when 40,440 persons were enumerated, the greatest percentage increase in Arizona's history occurred, almost 319 percent. Since 1880 Arizona has experienced an uninterrupted and rapid population increase. The population of Arizona

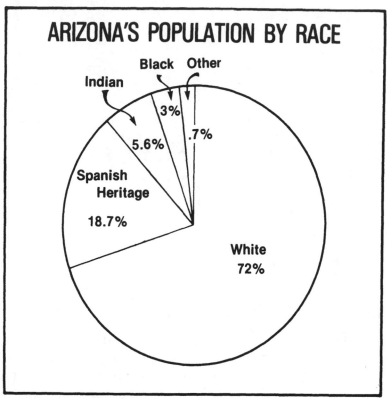

FIGURE 4.1
Source: United States Department of Commerce, Bureau of the Census.

FIGURE 4.2

ARIZONA POPULATION GROWTH

Year	Total Population	Persons/Square Mile
1864	4,573	.04
1870	9,658	.08
1880	40,440	.4
1890	88,243	.8
1900	122,931	1.1
1910	204,354	1.8
1920	334,162	2.9
1930	435,573	3.8
1940	499,261	4.4
1950	749,587	6.6
1960	1,302,161	11.4
1970	1,770,900	15.5
1978 (est.)	2,449,200	21.5

Source: U.S. Bureau of the Census. 1978 estimate from <u>Arizona Statistical Review</u>, 34th Annual Edition, Sept., 1978, Valley National Bank, Phoenix.

reached one million in the mid-1950s, but it took less than 20 years for a second million to be added, and Arizona is now well on the way to a third million.

As the population grew, greater and greater pressure was placed upon the environment. In 1960, about the beginning of the spectacular growth rate in the state, the population density was 11.4 persons per square mile (4.4/sq. km.) (Figure 4.2). By 1970 that figure had jumped to 15.5 (6), and by 1978 it had climbed to 21.5 (8.3). This figure is low in comparison to the national average, about 58 per square mile (22.4/sq. km.), but it is large for a rather fragile environment. Also, the 21.5 persons per square mile (8.3/sq. km.) are not even remotely scattered evenly around the state but are concentrated in the two large metropolitan areas (Map 4.1). This distribution is reflected in

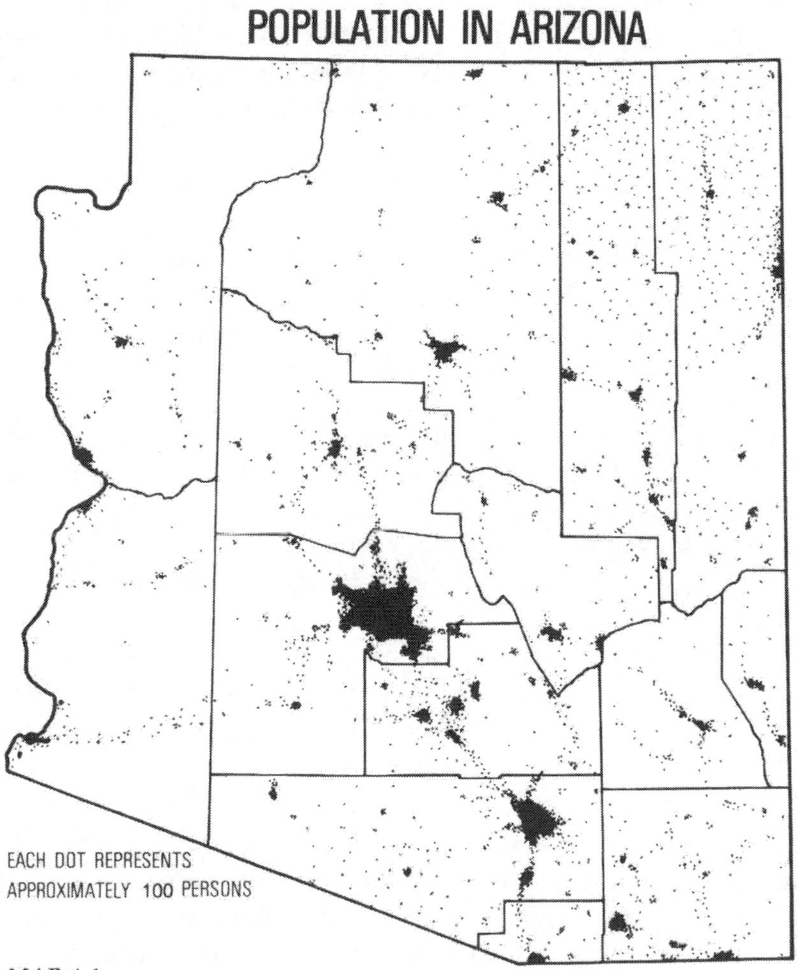

POPULATION IN ARIZONA

EACH DOT REPRESENTS
APPROXIMATELY 100 PERSONS

MAP 4.1
Source: U.S. Census, "Population, Urban and Rural in the United States,"
1970.

FIGURE 4.3

BIRTH RATE PER 1,000

Year	National	Arizona
1960	23.7	28.2
1975	14.8	17.9

Source: U.S. National Center for Health Statistics, Vital Statistics of the United States.

the fact that in 1970 Maricopa County had 105.7 persons per square mile (40.8/sq km.), and Pima had 38.1 (14.7), but Coconino County had only 2.6 persons per square mile (1/sq. km.), and Mohave only 2 (.8). This concentration of population puts great stress on the resources and recreational facilities near the two populous areas and along the highways leading to them.

Population growth in Arizona is the result of a combination of three factors: births, deaths, and migration. The natural increase of Arizona's population, the difference between births and deaths, is small in comparison to the growth due to migration. As an example, the natural increase of the population between 1970 and 1975 was 116,000, and the increase due to migration was 332,000. During the 1960s and 1970s, Arizona's birth rate declined although it remained somewhat higher than the national average (Figure 4.3). This lower birth rate means that Arizona would grow at a slower pace, except, of course, for the fact that so many people are migrating to the state. There is little hope that Arizona will achieve a zero population growth within the near or even distant future. Zero population growth is achieved when births equal deaths and immigration equals emigration, neither of which will happen soon in Arizona.

Most of Arizona's spectacular population growth has been the result of immigration. People seem to be moving to Arizona from every corner of the nation, as well as from Canada and Mexico. They are not moving there for only one reason, but for a wide variety of reasons: retirement, economic opportunity, a more relaxed life-style, the pleasant winters without snow, and many other factors. There is, of course, some emigration from the state. Many immigrants become disillusioned and either return to their home state or move elsewhere, and a significant number of Arizonans also emigrate — according to the 1970 census, one out of every three born in the state lives elsewhere. Nevertheless, immigration to the state far outstrips emigration.

About two out of every three present-day Arizonans were born elsewhere. In the 1970 census, over 1,100,000 Arizonans had been born outside the state, and the native born totaled only a little over 600,000; since that date immigration has greatly accelerated. The states that have contributed the greatest number of immigrants, not surprisingly, are the most populous ones; in order they are Illinois, Texas, California, Ohio, and New York (Map 4.2). Relative to their size (in terms of percentage), the states that have sent the most immigrants are the western north-central states, including Missouri, Iowa, Kansas, North Dakota, South Dakota, Nebraska, Minnesota, Wisconsin, and Michigan. These figures on Map 4.2 represent the states where the immigrants were born. Another interesting statistic is revealed by examining where those immigrants lived just prior to their move to Arizona, rather than where they were born. The five top states still lead the list, but their order is greatly changed. A 1971 study of the migrants to the Phoenix metropolitan area showed that 19 percent of them had come from California. The other top four states were Illinois (7 percent), Ohio (6 percent), Texas (6 percent), and New York (5 percent). It is obvious, therefore, that many people move to California, stay there for a time, and then move to Arizona.

Some parts of the state have been strongly influenced by immigration, but other areas have been largely untouched. Well over three-fourths of the migrants move to either the Phoenix or the Tucson metropolitan areas. Those areas of the state are therefore growing the most rapidly, and only a small proportion of their populations is native born, except among the young. Many migrants moving to those two areas are young people seeking employment, and, therefore, the birth rates there will be higher and the deaths fewer, which fuels the growth rate.

Although fewer people immigrated to some counties, such as Yavapai and Mohave, those counties were affected to an even greater degree because of the small original population base. Between 1970 and 1975, Mohave County experienced a net immigration rate of 41 percent, and Yavapai's was 32.5 percent. Those counties, therefore, experienced rapid growth even though the birth rates were extremely low. Births equaled deaths in Yavapai County, so there was no natural increase, and growth was the result of immigration. This situation also occurred in Mohave County where there were 900 more births than deaths, but the county increased by 11,500 within the five-year period. Most of the immigrants to Mohave County were elderly retirees, and those people can be expected to die within a few years so that deaths will soon be equaling births. Thus, Mohave County will probably not have any natural increase, but it can be expected to continue growing as other elderly people decide to retire there. As an illustration of the elderly makeup of those two counties, 10 percent of the state's population was 65 years or older in 1975, but in Mohave County, 12 percent was 65 or older, and in Yavapai 17.8 percent.

Some counties of the state have been largely untouched by immigration so far—this is true for Greenlee, Gila, and Yuma counties and, to a lesser ex-

BIRTHPLACES OF ARIZONA RESIDENTS, 1970

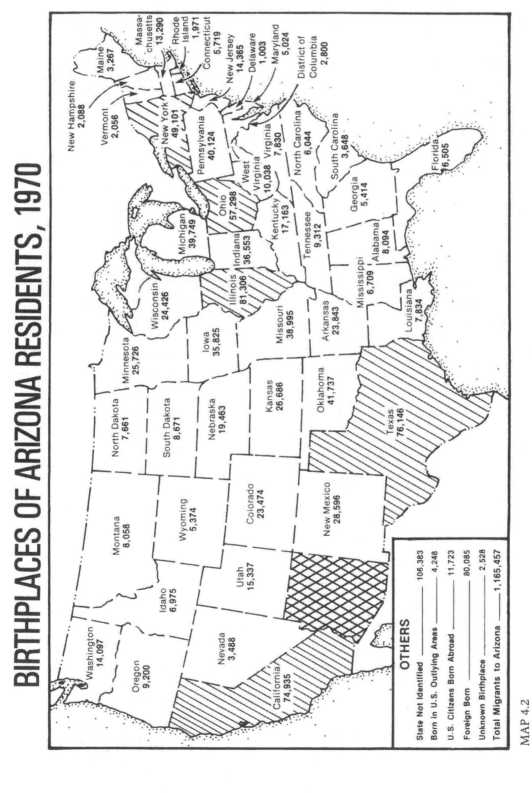

New Hampshire 2,088
Maine 3,267
Vermont 2,056
Massa-chusetts 13,290
Rhode Island 1,971
Connecticut 5,719
New Jersey 14,365
Delaware 1,003
Maryland 5,024
District of Columbia 2,800
New York 49,101
Pennsylvania 40,124
West Virginia 10,038
Virginia 7,830
North Carolina 6,044
South Carolina 3,648
Florida 6,505
Ohio 57,298
Kentucky 17,163
Tennessee 9,312
Georgia 5,414
Alabama 8,094
Michigan 39,749
Indiana 36,553
Illinois 81,306
Missouri 38,995
Arkansas 23,843
Mississippi 6,709
Louisiana 7,834
Wisconsin 24,426
Iowa 35,825
Minnesota 25,726
North Dakota 7,661
South Dakota 8,671
Nebraska 19,463
Kansas 26,686
Oklahoma 41,737
Texas 76,146
Montana 8,058
Wyoming 5,374
Colorado 23,474
New Mexico 28,596
Idaho 6,975
Utah 15,337
Washington 14,097
Oregon 9,200
Nevada 3,488
California 74,935

OTHERS

State Not Identified	106,383
Born in U.S. Outlying Areas	4,248
U.S. Citizens Born Abroad	11,723
Foreign Born	80,085
Unknown Birthplace	2,528
Total Migrants to Arizona	1,165,457

MAP 4.2
Source: U.S. Department of Commerce, Bureau of the Census.

tent, Navajo, Apache, and Cochise counties. A wide variety of reasons can be cited as to why immigrants have avoided these counties: much of the land is not open to settlement because it has been set aside as national forest, Indian reservations, and the like; there are few cities and employment opportunities; and there is a lack of certain amenities. The natural increase of the population in most of these counties has been relatively high, since few older people retire here and the population growth is largely a result of births. This is particularly true in those counties that have large Indian populations, for Indians have low death rates and high birth rates when compared with other groups.

The age-sex structure of the population is very similar to that of the United States (Figure 4.4). Many people think of Arizona's population as being "old" (having a disproportionate number of older people), but this is just not the case. Many older people do retire to the state, but the majority of the immigrants are 20 to 35 years old. Younger people are the most mobile, are willing to move to a new area for a fresh start, and also are within child-bearing years. Therefore, the national and Arizona age-sex pyramids reflect the fact that Arizona has relatively more young people and fewer old people than the nation as a whole (Figure 4.4).

There are, however, differences in the age-sex pyramids for the three major racial groups in the state. The age-sex pyramids for whites—which in this

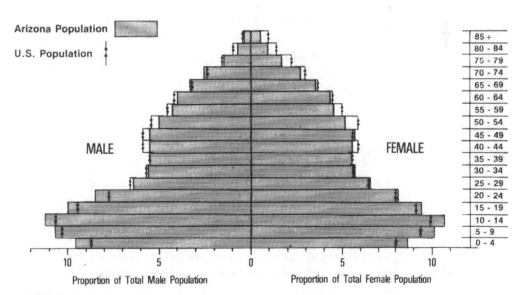

FIGURE 4.4
Source: United States Department of Commerce, Bureau of the Census.

case includes those of Spanish heritage—closely approximates the national pattern, but the black and Indian populations exhibit very different age-sex pyramids (Figures 4.5, 4.6, and 4.7). Age-sex pyramids for Indians and blacks indicate that those populations are young and growing rapidly. As an illustration, 36.3 percent of Arizona's population in 1975 was less than 20 years old, but more than 55 percent of the Indian population and almost 50 percent of the black population were younger than 20. It can therefore be expected that nonwhites will compose a larger percentage of the total population unless the immigration of whites continues at a high level. The Arizona Indian age-sex pyramid indicates a very young population. Indian births far exceed deaths, which results in a high natural increase. This is true in spite of the fact that the death rate for Indian children is much greater than the national average. The larger percentage of young remains because (1) the average Indian lives about 5.5 years less than the non-Indian, (2) the average Indian has little knowledge or concern for birth control (live births for Indians were almost double the national average—33.0 live births for each 1,000 Indians while the national average for all races is 17.3), and (3) the federal government has built many modern health facilities for the Indians. As more young people reach the reproductive age, the total Indian population will increase dramatically unless the fertility rate drops sharply or the death rate increases dramatically, and both are highly unlikely. Therefore, it can be expected that the Indian population will increase rapidly, at least to the end of this century.

The black population is also young, as is indicated by the age-sex structure. The indentation at the base of the pyramid indicates a reduction of the black birth rate, something not indicated as steadily for the Indians. It seems that the black age-sex pyramid may one day be more similar to that of the whites, but since such a large percentage of the black population will soon reach the reproductive age and since there are so few old people, the black population should increase for some time to come.

MORMONS IN ARIZONA

The Church of Jesus Christ of Latter-day Saints, commonly called the Mormon church, has its center in Utah, but its sphere of influence extends to Arizona, and Mormon history and influence have long been important to the state. Mormons cannot be called a minority group in a racial or ethnic sense, but they are, nevertheless, a distinctive subculture. Almost all Mormons are white and cannot be distinguished from the average white population of North America. They are very conscious of their subculture, which is based upon religion, and readily distinguish between the life-style of the members of the Church of Jesus Christ of Latter-day Saints (called Mormons, Saints, Latter-day Saints, and L.D.S.) from the life-styles of all non-

AGE-SEX STRUCTURE OF ARIZONA'S WHITES, 1975

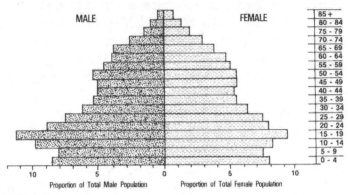

FIGURE 4.5
Source: United States Department of Commerce, Bureau of the Census.

AGE-SEX STRUCTURE OF ARIZONA'S INDIANS, 1975

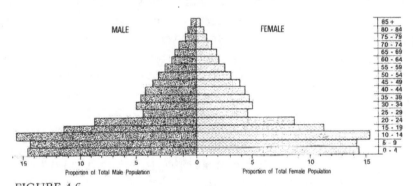

FIGURE 4.6
Source: United States Department of Commerce, Bureau of the Census.

AGE-SEX STRUCTURE OF ARIZONA'S BLACKS, 1975

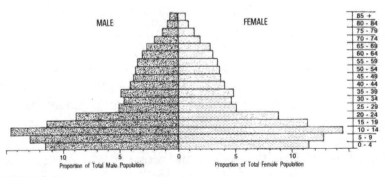

FIGURE 4.7
Source: United States Department of Commerce, Bureau of the Census.

Mormons (called Gentiles). The Mormon religion, and its subsequent subculture, have stirred strong emotions. There are many books, either for or
against Mormonism, that deal with Mormon history, culture, and beliefs,
but Mormonism is a difficult topic to cover, as is any religion, for people
believe in religion with their hearts and not with their reasoning minds.

The Mormons are worthy of study in connection with Arizona, not only
because there are 124,448 members in the state but also because they control much land in Arizona and have made a definite impact upon the landscape. Mormons have traditionally been farmers, or have lived in small
towns. They have often been among the first settlers in Arizona and the
West, and so they have owned and controlled a great deal of land. They have
shaped the land they own to fit their image of the ideal, and thus, in much of
the West a "Mormon landscape" can be found. A historical perspective is important, however, in understanding the Mormons and their values and
goals.

Joseph Smith, founder of the Mormon religion and one of nine children,
was born to a Vermont farmer in 1805. When he was in his early teens the
family moved to upstate New York. There was much religious excitement in
the early 1800s when Joseph was a youngster. This religious feeling was
strongest in the American South, but it was found all along the frontier.
Many small religious splinter groups were formed, and several completely
new movements evolved. It was an era when highly emotional religions
were particularly successful, and many groups used techniques such as tent
meetings and revivals to gain converts. The two most successful groups
were the Baptists and the Methodists. Joseph considered joining the
Methodists, and had he done so, the history of the West would have been
very different, but he decided to pray and seek God in his own way.

According to Mormon church history, Joseph had several visions and was
told the location of some golden plates. With the use of the "Urim and
Thummim" (two transparent "seer" stones set in bows of silver and attached
to a breastplate), Joseph transcribed the writing, and the result was the Book
of Mormon, first published in Palmyra, New York, in 1830. The church of
Jesus Christ of Latter-day Saints was founded at Fayette, New York, in that
same year.

Many "strange" religions sprang up away from the cities along the eastern
seaboard. Many were small sects of established religions, but others were
entirely new, such as the one established by Jemima Wilkinson in 1817. She
claimed she was the Christ and established a colony in Jerusalem, just 25
mi. (40.2 km.) south of Joseph's home at Palmyra. The Mormons were thus
considered another "different" religion, and since they, like other religious
groups, were being persecuted, Joseph decided to move westward to
Kirtland, Ohio (Map 4.3), a place chosen because another religious colony at
Kirtland (called the Campbellites or Disciples) embraced Mormonism. For a
time, the Mormons prospered in Ohio, but attacks resumed after the Panic

MORMON MIGRATION

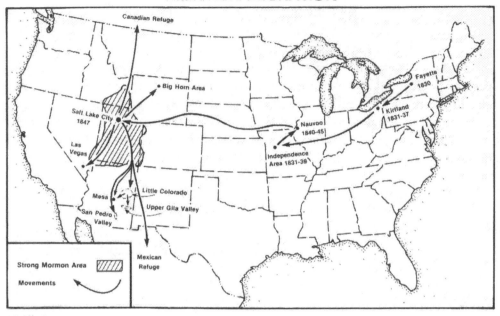

MAP 4.3

of 1837, and Smith decided again to move the group westward.

The place Smith chose as "Zion," the home for the Mormons, was Jackson County, Missouri, near Independence. Missouri was settled originally by southerners, who had more of a tradition of violence than did northern frontiersmen, and a strong hostility quickly developed between the Mormons and the Missourians. As soon as the first Mormons began arriving, violence erupted. Joseph Smith had a small army organized in Ohio to go to the aid of its coreligionists in Missouri, and the governor of Missouri threatened to call out the national guard. It was to be the bloodiest time in Mormon history, and for a long time afterward there were bad feelings between Mormons and Missourians. The hostility culminated in the Mountain Meadows Massacre in 1857 when a wagon train of about 140 persons, many of them Missourians, was attacked by Mormons and Indians in southern Utah and only a few children survived.

The Mormons quickly abandoned Missouri, and Smith, in 1840, secured a charter for a city in Illinois, Nauvoo. It was in this town that the Mormons first truly prospered, and within four years, there were 15,000 Mormons, 2,000 brick homes, and a large temple in Nauvoo, the most prosperous city in Illinois. Here, however, Smith allowed some Mormons to practice polygamy, and serious trouble resulted.

Smith controlled Nauvoo and the nearby country because the Mormons voted as a block. The interweaving of civil and religious affairs was familiar to a New Englander like Smith. To the Mormons (and the Spanish) only the crack of a gavel separated religious from civil functions. This system works well in places where there is only one religion and everyone belongs to it, but it engenders hard feelings on the part of those who do not belong, and this was the situation in Illinois. The real trouble began in 1844 when Joseph Smith sent the local marshal to destroy the press of a newspaper that had been attacking him. The burning of a press on the frontier was not a particularly unusual event, but it was the spark that was needed to arouse Gentiles in Illinois to action. Smith surrendered to the authorities and was taken to jail in a neighboring town. While there he was murdered by a mob of anti-Mormons.

Most people thought that the death of Joseph Smith would bring an end to the Mormon religion, but another strong and charismatic leader came forward, Brigham Young, who at the time was president of the Council of Twelve Apostles. He, too, was originally from Vermont, and he carried on the teachings and traditions of Smith.

Joseph Smith had contemplated a move further west, and Brigham Young decided that such a move had to be made. Young was familiar with the journal of John C. Fremont, who had explored the West in 1842, and with the book, *The Emigrants' Guide to Oregon and California,* by Lansford Hastings. Both mentioned the Great Salt Lake and the good grass, soil, and streams in that area, how isolated it was with towering mountains to the east and barren deserts to the north, south, and west. Some Mormons who had been to California tried to convince Young to move to the San Joaquin Valley, but he was convinced that the Great Salt Lake area was to be their home. The move to Utah began in 1846, and by 1847 the first Mormon settlers were arriving in the Salt Lake area.

The Mormons developed a strong base in Utah and from there began extending their influence into the Arizona area. There were four reasons the Mormons began moving to Arizona: (1) a desire to work with the Indians, converting them to Mormonism and, as the Spanish before them, raising their life-style, both materially and socially; (2) to relieve population pressure in Utah; (3) to try to return to a social system called the United Order; and (4) to find a refuge for the practice of polygamy.

The first Mormon intrusions into Arizona from Utah were in an attempt to work with the Indians. The Mormons had heard that the Hopi practiced agriculture and had some semblance of civilization. This made the Hopi very different from the Indians in Utah, and the Mormons thought that perhaps the Hopi were remnants of the ancient civilizations mentioned in the Book of Mormon. Jacob Hamblin was the man who was "called" to work among the Hopi. (A "call" was an official selection by the church hierarchy – who were considered prophets of God – and thus the call was considered divine, and acceptance was made accordingly.) Hamblin and his party left southwestern Utah and after three days' travel, arrived at Pipe Springs, now a national monument. They crossed the Colorado River at the only known crossing at the time, "the Crossing of the Fathers" – the site was originally used by the Spanish padres, hence the name (it is in Utah and is now covered by the waters of Lake Powell).

The Mormons found the Hopi very different from the Kaibab, Paiute, and other Indians of Utah. The Hopi men worked in the fields and at looms, whereas most Utah male Indians would only hunt rabbits and considered agriculture the work of women. The Mormons were welcomed by the Hopi, for they had a legend that white men would come from the west. The Mormons were also fortunate in that a local chief, Tuba, became their friend and supported their cause. They made many visits to the Hopi, especially Jacob Hamblin who went on several occasions, but the Mormons, as the Spanish before them, met with little lasting success.

Some good, however, did come to the Mormons from this contact. Much of northern Arizona became known to the Mormons, and some small settlements were started. Through their association with Arizona and their work with John Wesley Powell, the Mormons managed to have pressure put on the Navajo to stop their raids on Mormon settlements. Finally, and most important, the crossing of the Colorado at Lee's Ferry was discovered and exploited.

Lee's Ferry became very important for Mormon expansion southward. The site was named for John D. Lee, who lived at this isolated spot because he was wanted by the federal authorities for his involvement in the Mountain Meadows Massacre. (For his role in that affair, he was eventually arrested, convicted, and executed.) The site of Lee's Ferry was just above Marble Canyon where the Paria River meets the Colorado, though the actual crossing was about a half mile above the junction (Map 4.4.). This was an easy and safe crossing, but once the Mormons had crossed the river, the most difficult portion of the whole trip was before them because the travelers were forced to ascend a steep and rough ledge known as Lee's Backbone. The Mormons built one road directly across the Backbone and one around it, but both remained arduous and difficult. A third route was developed and it involved a dugway built along the edge of a cliff that reached the level of the valley floor south of Lee's Backbone. (It was the best route and remained in use until a bridge was built across Marble Canyon in the late 1920s.) Another ferry site was below the juncture of the Colorado and Paria rivers, but it was not a good site as there was little room to land. This latter site was at the very head of Marble Canyon, and if one were swept into that canyon, there was little chance of escape. In spite of Lee's Backbone, the site remained the only good crossing of the Colorado in Arizona for a long time, and it was widely used by Mormons traveling back and forth between Arizona and Utah. Other routes, however, were also used (Map 4.5).

GENERALIZED MAP OF LEE'S FERRY

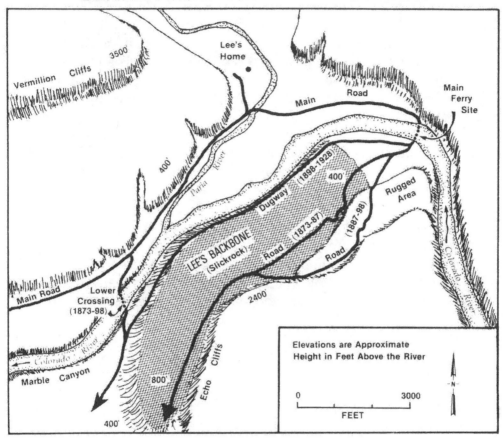

MAP 4.4

Population pressure was another reason for the Mormon movement southward. In 1850, Utah had about 11,000 residents, but by 1870, the population had risen to approximately 86,000. Although the birth rate was high, a great deal of the increase was a result of immigration, and much of that was from Europe. Between 1860 and 1870, about 20,000 European immigrants arrived, well over half of them from the Scandinavian countries. By 1870, one-third of the residents of Utah were foreign born—a higher percentage than in any other state or territory in the Union. Few Mormons in this early period were miners or even ranchers; the best way to hold the land was to farm it, so farming was encouraged by the church. There are many western novels concerning the conflict between the cattlemen and the sodbusters; in real life, the sodbuster was often a Mormon. There was a scarcity of water for agriculture in Utah, however, and soon the Mormons were having to look elsewhere to establish agricultural communities. Most of their expansion was northward, but they also expanded southward to present-day Arizona.

MORMON ROUTEWAYS TO ARIZONA

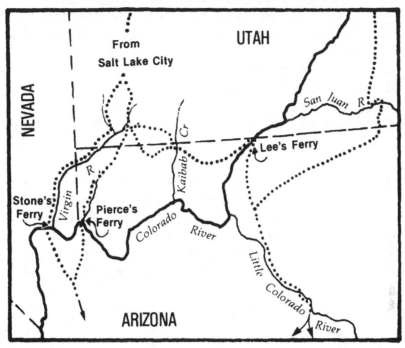

MAP 4.5
Source: Charles S. Peterson, "Settlements on the Little Colorado, 1873-1900." Unpublished Ph.D. Dissertation, University of Utah, 1967.

A third explanation for the Mormon expansion into Arizona was the desire to return to the social system known as the United Order, which had been tried in the decade prior to the death of Brigham Young in 1877. The United Order resulted from a revelation Joseph Smith received from God in 1831. The Law of Consecration and Stewardship was based on the idea of equality and cooperation between members and on the concept that the land was placed on earth by God and that humans were placed here only as caretakers of the land's abundance. The principle was developed very early and had much meaning, but the Mormons apparently never took it too seriously because they never really practiced it. They did attempt to follow the ideas of the United Order at Kirtland but did so in only a modified form in Missouri and at Nauvoo and very little in Utah.

The United Order was a social system whereby there was complete cooperation and sharing within the group. Property, crops, animals, and other such items were held in common, and only individual homes and some personal property, such as furniture and clothes, were privately owned. Meals were served in a community dining hall, there was a nursery for child care, and washing was done in a town laundry. There were "department heads" who were responsible for the different functions—for example, one person was responsible for the cattle and another was put in charge of the laundry – and each had to make a report to the total group at town meetings. In many ways, these settlements resembled an Israeli kibbutz.

Many towns were founded along United Order lines, but few succeeded for any length of time. Examples of such towns in Utah are Orderville, considered a success and a model for others, and Brigham City. United Order settlements were established in Arizona, but none lasted for more than a year or two. The only way a United Order settlement could be successful was if it was in a rural setting, if everyone was Mormon, and if everyone was willing to cooperate.

A final reason for Mormon expansion was because of problems caused by polygamy. Polygamy was first practiced in Nauvoo and was openly practiced in Utah, but the American government was determined to end the practice. The army established a large military base just outside Salt Lake City; it could be seen from the town, and it reminded the Mormons that the government was serious in its determination. The threat caused a consolidation of church forces, and some areas were abandoned as the men were needed for possible defense of the home area. Also, polygamy caused expansion because the polygamous Mormons were always searching for an isolated spot in which to live and practice their way of life, undisturbed by government intervention.

Not all Mormons practiced polygamy – indeed, according to the records of the Mormon church, only 3 percent of the members were polygamous – but the concept was supported by all. The church thought the government had no right to intervene in the church's affairs. Polygamy, to the Mormons, was not simply a social institution to be legislated away by Congress, it was a theological doctrine instituted by God through his prophet Joseph Smith. Since the United States was based on the idea of a separation of church and state, the state had no right to interfere. The average American, however, looked at the situation very differently and was determined that the government should end the practice of polygamy. The Mormons relented in 1890; they said a revelation had been received from God, and they issued a manifesto to end the practice of polygamy.

But polygamy still exists in two ways. Some splinter groups broke away from the Mormon church and still practice polygamy. Thus, it is surreptitiously practiced in Arizona, particularly in the northwesternmost corner of the state. The other way polygamy exists in the minds of the Mormons is in the afterlife. To a Gentile, a marriage lasts until "death do you part," but not to a Mormon, as a Mormon marriage "sealed" in a temple is for eternity. If a wife should die before her husband, he could be married and sealed to a second wife. He would not be a bigamist according to the laws of the land, but he then believes he will have two wives in the next life.

The first large Mormon settlement in Arizona was along the central reaches of the Little Colorado River. The first expedition that attempted to settle in that area returned to Utah after they found no suitable site. Brigham

Young sent an explorer to examine the territory, and he returned with a glowing account about the agricultural potentials along the Little Colorado, so more settlers were sent. The second attempt at settlement, organized along United Order lines, was in 1876, but it too was a failure and was abandoned by 1878. The only town to survive was Joseph City, and it has survived only because it was between the successful railroad towns of Winslow and Holbrook. The settlements failed because of poor soil, harsh climate, and especially water problems. Dams were repeatedly built and washed away by floods; the Little Colorado was just too big a stream for the early settlers to control, given their technology and resources.

With abandonment of the settlements along the Little Colorado, most of the settlers moved southward toward the Mogollon Rim and the upper reaches of the Little Colorado and its tributaries. To support themselves the first winter in this new area, many of the men worked for the Santa Fe Railroad, which was then being built across the territory, and they were greatly aided by the fact that the contractor for much of the work was John Young, son of Brigham. Mormons were not the first settlers in this area, and generally they had to buy their land and water rights. The people who initially controlled the land were Gentile cattlemen who had moved westward from New Mexico. They were mostly Texans but employed Hispanos from New Mexico as laborers. The Mormons established many towns in this area in the late 1870s, such as Springerville, Heber, and Show Low (named after a card game), Snowflake (named for two men), and Taylor. St. Johns is an exception, for it initially was an Hispanic town. However, it later became rather evenly divided between Mormons and Hispanic Catholics, and the entire area is strongly Mormon in character (Map 4.6).

The Mormons soon expanded into southern Arizona. There were several reasons for this southward movement, such as the reports of rich land available in the Salt River Valley and the desire to establish settlements there in preparation for expansion into Mexico. Lehi, established in 1877 as Utahville, was the first Mormon settlement in the Salt River Valley. The settlement failed, partly because of physical problems, but especially because of internal dissension – in spite of the fact that it was a United Order settlement. A year later Mesa was established just 3 mi. (4.8 km.) southwest of the Lehi townsite, mostly by people coming from Utah, though some came from the Little Colorado settlements. Mesa was a successful settlement, even though the Mormon settlers had a difficult first year. They were greatly aided in their survival by working for Charles Trumbull Hayden, a Gentile who wisely encouraged the Mormon settlement. He had a flour mill, and the Mormons were good farmers who could grow a large amount of wheat; it was a mutually advantageous situation.

The Mormons also expanded into two other areas of southern Arizona (Map 4.3). One small but very successful settlement was in the San Pedro River valley, centered around St. David. The farmers in that area sold their products in the mining camps of Tombstone and Bisbee, in the railroad

SETTLEMENTS NEAR THE LITTLE COLORADO AND MOGOLLON RIM

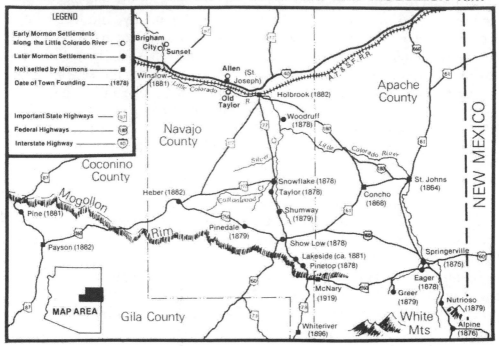

MAP 4.6

town of Benson, and to the nearby military posts. The other area of Mormon settlement was along the upper Gila River, particularly around the town of Safford.

Several generalizations can be made about the Mormon settlements. In all of them, the Mormons were farmers, or sodbusters, who acquired the water rights, but the Mormons were the initial settlers in only a few of the areas. Preceding them were miners, traders, ranchers, and agriculturists. In some areas, such as around Mesa, the land and water were for the taking, but in many other instances it had to be bought from earlier settlers. Small Mormon enclaves developed, and they were surrounded by Gentiles. The Mormons were basically small, independent farmers. Their settlements were generally homogeneous, since the Mormons were almost all of north European ancestry, and they employed very few blacks or Mexican Americans on their farms. Adding to the uniformity of these settlements was the fact that almost all the settlers had lived in Utah for several years and thus had gone through a "Utah experience" before moving to Arizona.

The Mormons eventually spread into Mexico, mostly into the state of Chihuahua, but some spilled over into Sonora (Map 4.7). The reason for the expansion into Mexico was, again, to find a refuge for the practice of

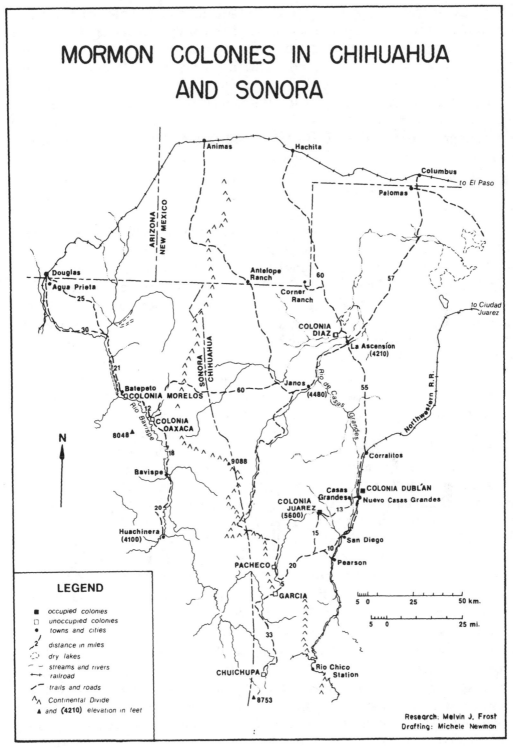

MAP 4.7
Courtesy, Melvin J. Frost.

polygamy (there was also expansion into Canada for the same reason), and almost all of the families in the colonies in Mexico were polygamous. These Mormon settlers literally "caught hell" during the various Mexican revolutions. They wanted to remain neutral, but they were usually abused by both the revolutionaries and the federal troops. They were also greatly resented by the local Mexicans, who would often squat on Mormon land—the last such event occurred in 1977. The Mormon colonies in Mexico slowly began to disintegrate, especially after an 1893 manifesto prohibited polygamy and gave amnesty to all who had become polygamists prior to 1890. Today, only two Mormon colonies in Mexico survive, Colonia Juárez and Colonia Dublán, consisting of a total of about 600 to 700 people who primarily produce apples for sale in Mexico City.

The Mormons settled a harsh environment in the West. It was an unforgiving environment, and one with which they were completely unfamiliar. The Mormons came from the humid East or northern Europe, and they knew nothing of deserts, local soils, or the local climate, and certainly they knew nothing of irrigation techniques. But the Mormons adjusted to their new environment very quickly, and within a few years irrigated agriculture was a success. The main reasons for the Mormon successes were probably the close cooperation of all the settlers and the control, guidance, and support given by the church. As farmers who controlled a great deal, if not most, of the agricultural land of the intermountain West, they made a big impact on the land as they changed it to fit their goals and ideals. The Mormons developed certain ideas and traits that they impressed on the landscape, and these traits identify what is termed a "typical Mormon landscape."

One researcher has identified 10 characteristics of the typical Mormon landscape in the West (R. V. Francaviglia, "The Mormon Landscape," *Proceedings of the Association of American Geographers* 2 [1970], pp. 59–61). Most of these Mormon traits are found in Arizona, and if several are seen in one community, it was undoubtedly settled by Mormons. Many of these traits are changing, but they still can be commonly seen and recognized in rural areas. The 10 characteristics are (1) village life, (2) wide streets, (3) roadside irrigation ditches, (4) outbuildings in towns, (5) unpainted farm buildings, (6) the "Mormon fence," (7) architectual styles of houses, (8) brick-and-stone houses, (9) ward chapels, and (10) a large number of trees.

Village life is a trait commonly recognized in the West as being Mormon. A traveler going westward through the farming areas of Kansas and Nebraska will find farms widely scattered on the land, but in Utah the traveler will notice towns at regular intervals along the agricultural valleys, each about 10 to 12 mi. (16 to 19 km.) apart. Between these towns, there will be cultivated fields and well-kept barns, but few houses because the farmers live in the towns. It was Joseph Smith's plan that all should live in towns, even farmers. This tradition can be traced back to Smith's roots, for in pioneer New England most of the farmers lived in towns, and this settlement pattern is also common in Europe, so most of the Mormon immigrants from Europe were comfortable with the system. The idea of farmers commuting to their fields is certainly not typical in the United States, but it was a concept the early Mormons maintained, and only now is it changing.

Most of the Mormon attitudes about towns and cities come from the plan for the "City of Zion," which was drawn by Smith while he was living in Ohio. It was to be the plan for the ideal and perfect city, and the ideas presented in the City of Zion plan were to permeate Mormon city building throughout the nineteenth century. But apparently it was something the Mormons never considered as divinely received from God, for they never followed the plan exactly. A City of Zion was to cover 1 sq. mi. (2.6 sq. km.) and be divided into 42 blocks; each block would cover 10 acres (4 ha.), and there would be 20 lots to each block. The barns and stables were to be to the north and south of the city. Agricultural land was also to be north and south of the city, but after farmers had to go too far to reach their fields, agricultural land could then be developed to the east and west. Smith believed that from 15,000 to 20,000 inhabitants would live in such a city, but in reality such a town could hold only about 5,000 people.

The first place the Mormons extensively used and modified the City of Zion plan was in Nauvoo. The changes that were made there resulted in what is called the "Nauvoo modification," and it is the plan most commonly followed in Mormon towns in the West. In Nauvoo, there were 4 lots on each block of 5 acres (2 ha.), and each villager received a farm of 20 acres (8.1 ha.) and a share in a communal pasture. This division of land was commonly followed in the West, and Snowflake is an example of a settlement that closely parallels this pattern. Twenty acres (8.1 ha.) was not enough land to maintain a farmer and his family, so the farmers also acquired land wherever it could be obtained, and the result was scattered field ownership.

The second characteristic of the Mormon landscape is wide streets in the towns (Figure 4.8). Most Mormons realize that their towns have wider streets than Gentile towns, but they do not know why. When asked why the streets are so wide, they will usually answer, "when the town was built, they had to be wide enough to allow an ox team to turn around." Actually, the reason is because the City of Zion plan called for all streets to be 132 ft. (40.2 m.) wide. Smith was probably influenced by the fact that in Cleveland, which was close to Nauvoo, all major streets were exactly that width. Since then, the Mormon towns have had wide streets. However, the Mormons never took the figure of 132 as divine, for most of their towns have streets of 80 to 120 ft. (24.4 to 36.6 m.) wide. Each Mormon town today, no matter how small, seems to have been planned as a metropolis because of its broad streets. Often just the centers of the streets are used, and 30 to 50 ft. (9.1 to 15.2 m.) of grass and weeds extend on either side.

The third characteristic is the roadside ditches. Along the wide streets there are small, shallow irrigation ditches, which often resemble the deep gutters in non-Mormon towns. The roadside ditches usually appear as weed-filled slots, but they can be recognized as irrigation ditches by the headgates.

FIGURE 4.8. A wide street in Mesa, Arizona—a Mormon town.

The building of these canals was an integral part of town planning in the Mormon settlements in the West. They were critically important for use in watering the town dwellers' orchards and gardens, which were called for in the City of Zion plan.

Another characteristic of the Mormon landscape is the large numbers of barns and other outbuildings in the towns. Many of the town lots are large, and portions of them are devoted to farm use, including outbuildings such as stock barns, hay barns, sheds, and granaries. This tradition of having barns inside the towns developed in Nauvoo, where it was realized that barns outside of town would be threatened by vandals.

The fifth characteristic is the unpainted farm buildings. Barns are usually painted red, white, or some other color in non-Mormon areas, but a large number of unpainted barns in a village or a dispersed settlement indicates a Mormon community. The Mormons even joke about their unpainted barns, but just why the tradition began is unknown.

The "Mormon fence" is another common landscape feature. It is a crude, unpainted fence made of whatever the farmer has handy, including pickets, planks, and old car hoods (Figure 4.9). This type of fence originated prior to the invention of barbed wire and during a time of open-range ranching when the farmers were often at the mercy of the ranchers. This kind of fence still survives in a few areas, especially around farmsteads and barns. As the name implies, local people recognize it as a Mormon fence.

The seventh characteristic is the two distinctive architectural styles for the houses. For a long time, the Mormons were very conservative and re-

jected any new fads and styles in architecture. As a result, the two traditional house styles are still commonly seen in Mormon areas. One style of
house associated with the Mormons is the "I" house. The I-house tradition
probably began in Nauvoo, for it is a midwestern style. The typical I-house
has two stories, a central hall, chimneys at either end, and a symmetrical
facade (Figure 4.10). The basic design is only one room deep and two rooms
wide, but all such houses have been modified and have additions in the rear.
Another characteristic of this house is the use of dormers, a Scandinavian
tradition. The Mormons have many names for this style of house, such as
Nauvoo, Mormon, or Polygamy house. A second Mormon house style in
Arizona is the "square" home (Figure 4.11). The basic design for this type of
house calls for four rooms, one story, a hipped or pointed roof, and a symmetrical facade. One great disadvantage of this type of home is the difficulty
of adding rooms; such a house is usually enlarged by building on small
"sheds" (by extending the roof on the sides) or by enclosing the porches.
These traditional houses are no longer built by the Mormons, who today
build modern and conventional ranch-style homes, but the older houses are
very visible and are still occupied.

Building in brick or stone is another Mormon characteristic. Any town
west of the Rockies with over half the houses built of brick or stone, is likely
to be a Mormon town. It is not a New England or midwestern tradition, for
there people built structures with wood. Rather, it goes back to religious doctrine, for brick or stone homes were called for in the plan for the City of Zion.

The ninth characteristic of a Mormon landscape in Arizona is the ward
chapel. These chapels are usually located in the center of a town, and they
are always of brick or stone. The typical ward chapel has been described as
"'colonial style' with heavy Greek revival pediment and cornices, or a more
modern abstraction of that form" (Francaviglia, p. 60). All of them lack
originality, as the Church Architect Office in Salt Lake City has several basic
designs of which one must be chosen by a local congregation. The ward
chapels contrast sharply with Protestant churches, for they have no crosses,
only spires. The ward chapels are also larger because they are used for social
activities throughout the week.

A final characteristic of a Mormon community is a large number of trees,
because one of the first things the Mormons did when they settled was to
plant trees. The Lombardy poplar is particularly associated with Mormons,
and that tree is commonly found in yards and cemeteries and planted to
form windbreaks around the fields. The Italian cypress, which is similar in
appearance, is commonly substituted for the Lombardy poplar in southern
Arizona. The Mormons also planted many fruit trees around their homes.
Because of the tree plantings, a Mormon community would have a "settled"
look in just a few years—something a Gentile community might have taken
25 years to achieve.

Yet another Mormon trait mentioned by Francaviglia is the hay derrick,
sometimes called the Mormon stacker or Mormon derrick (Figure 4.12). It

FIGURE 4.9. A "Mormon fence" on a farmstead in northern Arizona.

FIGURE 4.10. A typical I house. Note the chimneys on either side, the dormers on the roof, and the extension at the rear.

FIGURE 4.11. A square home in Mesa, Arizona.

FIGURE 4.12. A Mormon stacker in the Sevier River valley in southwestern Utah. These stackers are still commonly seen in Utah, although today they are seldom used. The large use of wood in the fence is still another trait often found in Mormon agricultural landscapes.

consists of a long, vertical wooden post with a horizontal beam near the top, and it is particularly used to stack the hay into haystacks. These hay derricks are found throughout Mormon country wherever alfalfa is grown – except for Arizona. For some strange reason, it was apparently never introduced to that state, in spite of the facts that alfalfa is grown there and all early Mormons came from Utah where they were familiar with the hay derrick.

Other characteristics may be added to those listed above, but the first 10 are the most important clues in identifying the Mormon landscape in Arizona. If any town or region has most of the above characteristics, it will undoubtedly be Mormon. The typical Mormon landscape is the result of over 75 years of Mormon occupancy and control. Very recent Mormon settlements are more difficult to recognize, but certain subtle clues still persist.

MEXICAN AMERICANS

Prior to the Mexican-American War, the only Mexican citizens living in present-day Arizona were in the Santa Cruz Valley, from Tucson southward. The population had increased there, but Apache troubles became severe after 1810 and later civil disorders occurred, and all of Sonora lost population, including the Santa Cruz Valley. By 1835, ranching had ceased, and there was only a little farming near Tucson, a town of about 600, and Tubac, with a population of about 400. Thus, at the time of the Treaty of Guadalupe Hidalgo, there were no Mexicans in the northern half of what was to be Arizona, and in the territory acquired by the Gadsden Purchase there were only approximately 1,000 in the towns of Tucson and Tubac (Figure 4.13).

Treaties between the two governments guaranteed the rights of Mexicans living on the northern side of the border. The Treaty of Guadalupe Hidalgo, for example, stipulated that "property of every kind, now belonging to Mexicans not established there, shall be inviolably respected." It is hard to tell what was happening to the population at that time, because it was an unstable era. Some believe that many Mexicans fled the Santa Cruz Valley because of uneasiness and fear, and others contend there was a flood of Mexican citizens to Arizona Territory to take advantage of the fact that they could gain U.S. citizenship if they remained in the area one year. Other reasons for moving north were greater economic opportunity in the United States and greater protection from Apache depredations.

The Mexicans and local Mexican Americans were particularly important in the mining industry during the pre–Civil War era. Many Americans began opening mines in the Gadsden Purchase area, around Tubac and east toward Patagonia, and they employed mostly Mexican labor. Many of the Americans who had gone to California in the gold rush days began filtering into Arizona in search of riches. Many Mexicans also came to Arizona in the

FIGURE 4.13. Tucson about 1869. Source: John Ross Browne, *Adventures in the Apache Country* (New York: Harper and Brothers, 1969).

same way, by first working in the California gold fields and then drifting into Arizona to work the mines there. Many began to work the placer gold deposits along the lower Colorado and Gila rivers, and for the first time that area developed a permanent Mexican-American population. There was soon a growing economy in southern Arizona, and Tubac had a population of about 1,000 persons, most of whom were Mexicans or Mexican Americans.

With the outbreak of the Civil War, mining activities quickly ground to a halt, and the Apache began to pose a real threat. A little placer mining continued in the southwest and around the Bradshaws, and some Mexicans came north to illegally mine ore, but that activity was of little consequence to the Mexican Americans in southern Arizona, and few Mexicans moved north at that time.

Mining began again in earnest a few years after the Civil War, but there were now many changes in the industry. Mechanization began to assume an importance, and there was little need for unskilled Mexican labor. Also, much of the mining activity shifted to the Mountain District of Arizona, and that area developed as a predominantly Anglo-American region with strong Mexican-American prejudices; for example, Globe became a center of anti-Mexican activity. One exception was the Clifton-Morenci area. Mexican labor had been recruited at El Paso to work the mine in that area, and it was always called a "Mexican camp." With the passage of time, more and more Mexican laborers filtered into the mining camps of the mountains. Many went originally as strikebreakers, and that fact accounted for much resentment, but with time, they were accepted, and eventually almost all mining towns had a large Mexican-American population.

Though the mining activity receives much attention, most Mexicans after 1870 began finding employment as agricultural workers. Most agricultural development was in the Salt and Gila valleys, and Mexicans began migrating there, so that by the late 1870s, about half of Phoenix and Florence populations were Spanish speaking. It was not very long, however, before mechanization began to displace farm laborers in Arizona.

Mexican Americans have owned very little land in the area since it was acquired by the United States, and thus their impact upon the environment has been relatively slight. The Mexicans who had money stayed in Mexico, or if they did come north of the border, they did not invest in land. The majority of the Mexicans who migrated north of the border were illiterate and impoverished, and they could not afford to buy land. Farms could be acquired at very little cost, through homesteading and other means, but most Mexicans could not take advantage of those opportunities because they did not know about the laws and many were not citizens. Although many were agricultural laborers, they lived in towns and villages and commuted to the fields, a tradition that continues to this day (Figure 4.14).

FIGURE 4.14. Buses used in taking agricultural workers from towns to the fields. Note the portable toilet behind the bus.

There were two major reasons Mexicans came north, economics and politics. The economy on the U.S. side of the border has been much more dynamic, and that fact has lured many Mexicans to abandon their home country in hopes of earning a better living for themselves and their families. The lack of political stability in Mexico will invariably cause migrants to move north, especially if the unrest should lead to a violent revolution. Any revolutionary era in Mexico has caused a flood of political refugees to move north, and the situation has not changed. If and when there is another revolution in Mexico, political refugees will once again move northward.

Railroads both directly and indirectly influenced Mexican migration into Arizona. Mexicans were not important in railroad construction, because most of that manual labor was done by others, especially the Chinese. But the railroads indirectly affected the Mexican population because the railroads provided cheaper transport, which meant that copper mining could be successful. Many copper mines developed in Arizona and Mexico, and many Mexicans were employed in those mines. Railroad construction in Mexico in the early 1900s brought many changes to that country. Railroads paid good wages, and mining employment boomed. The haciendas soon began to lose their peons—those persons tied to their employers because of debts—because the peons began fleeing to find better economic opportunity elsewhere. In fact, entire villages were abandoned. The northern terminus of the railroad was El Paso, and so that city became a great magnet for Mexicans. El Paso became a clearinghouse for Mexican labor, and gangs of laborers were organized there and then sent to jobs throughout the West. Many Mexicans entering the United States, especially after 1900, were coming to the United States from deep within central Mexico—many even came from south of Mexico City. Soon there was a critical labor shortage in Mexico, and high wages were demanded by those who had remained. The problem was partially solved by importing Orientals into Mexico, many of whom have descendants there.

The situation in the Pueblo area of New Mexico was quite different from that in Arizona. The Pueblo area of New Mexico was culturally diverse at the time of American acquisition. There were Indians who had maintained their cultural identity, largely unchanged by the Spanish newcomers, and there were some people of pure Spanish blood, such as a few religious and some soldiers and administrators. In 1800 there were approximately 20,000 "Spanish" in the pueblo area, but the vast majority of those were mestizos, or a mixture of Spanish and Indian. Many Spanish and mestizos had moved to the area and had intermarried with local Indians who accepted the Spanish language, religion, dress, and other aspects of the culture. With time this group, known as Spanish-Americans or Hispanos, became very powerful politically and economically. Although overwhelmed by the Anglo immigrants who have entered the area since World War II, the Hispanos remain a powerful political and economic force. There is no similar large and powerful Hispanic group in Arizona.

Although the Mexicans have greatly expanded northward, the Hispanos of New Mexico have seemed content to remain in the area in which they first settled. Because of Indian pressure, the Hispanos did not move outward until after the beginning of the American period, and by that time, they were forced to compete with Texans and other Americans for any new land. Therefore, they managed to dominate few new areas. The only town settled by these Hispanos in Arizona is St. Johns, an old Hispanic town, but since the 1870s even St. Johns has been considered to be about half Hispano and half Mormon. It would seem logical that the Hispanic population of north-central New Mexico, with its Spanish culture and tradition, would attract many Mexicans to the area, but that has not been the case. Mexicans moving north have tried to move to areas of potential employment, and the Hispanos usually provide the cheaper labor for that area, so it has been avoided by Mexican migrants.

The Mexican-American population in the Southwest has grown rapidly since the beginning of the American period, but that group is difficult to count, and the figures are not particularly trustworthy. In the earliest American period, the Mexican Americans were easy to identify because most were born in Mexico. Between 1870 and 1880, for example, the Mexican-born element of the population in Arizona more than doubled from 4,348 to 9,330. In 1850, soon after the American acquisition of most of the Southwest, there were about 80,000 Mexican Americans in the region (Texas to California). By that time, however, they were already a minority, for the total population was 366,600, and that figure did not include the nomadic Indians.

The growth of the Mexican and Mexican-American population in the Southwest was spectacular after 1900. There were only a little over 100,000 Mexican Americans in the Southwest in 1900, but by 1930 that figure had grown to 1,250,000. At the time, probably one out of every 10 Mexican citizens was living in the United States. There were several reasons for this phenomenal growth. First, the great Mexican revolution from 1910 to 1917 caused many to leave Mexico. Second, there was great material poverty for most Mexicans. Third, World War I created a demand for labor in the United States, and, fourth, the booming war economy continued into the 1920s. The U.S. government did little to discourage this migration, but the depression of the 1930s changed that policy, and many Mexicans were forced to leave. The expanding economy experienced by the United States since World War II has again attracted Mexicans to settle in this country for economic reasons, and even small recessions in the economy have not reduced their numbers because they usually work at the most menial jobs and for the lowest wages.

Many of the migrants moving north did so without bothering to go through customs and were therefore considered "illegal aliens"; hence, they had few legal rights, and many were exploited by employers. The worst abused were the farm laborers. When it came time to be paid, many would find that their employers, instead of paying them, would have called the U.S. Border Patrol to take them back to Mexico. Because of this and other abuses, the U.S. and Mexican governments concluded an initial treaty in 1942 whereby Mexican nationals could legally work in the United States. This program went through several changes, but as a result of domestic and Mexican demands, Congress in 1951 established Public Law 78, which became known as the Bracero program. This program was established to control the flow of labor, to give the aliens legal status, and to provide them with decent housing, insurance, and a minimum wage. By the late 1950s, over 400,000 temporary Mexican agricultural workers were admitted into the United States annually. Although the Bracero program helped the economies of the United States and Mexico, it had a depressing effect on farm workers' wages. Domestic agricultural workers were affected, and unionization could never succeed so long as such cheap labor could be acquired so easily. The program was extended and changed with time, but it was finally ended in 1964, and only then could the condition of the American farm laborer improve and unionization begin. A few temporary workers still cross the border under what has been nicknamed the Green Card program (Public Law 414), which allows foreign workers to work in the United States if they have a sponsor who guarantees them work, either daily or for a longer period of time. However, the number of workers entering the United States under this program are few in comparison to the number that entered under the Bracero program.

Mexicans are still entering the United States illegally. It has recently been estimated that there are about 6 million Mexican citizens illegally in this country, almost 9 percent of Mexico's total population and an even higher percentage of Mexico's work force (and some estimates of the Mexican population living in the United States are as high as 10 million). Present conditions indicate that the situation will not get better, for Mexico has maintained a steady population growth of 6 percent over the past few years—one of the highest rates in the world. This population growth rate is coupled with slow economic growth and a high and growing rate of unemployment, a rate estimated, if underemployment is included, at 40 percent. Also, the income gap between rich and poor has increased, which has only worsened an already bad situation. Therefore, it can be expected that more and more impoverished Mexicans will move north, because they can see little chance for improving conditions in their own country. Many of these Mexicans migrate deep into the United States, especially to the large interior cities. Today smugglers probably bring in the majority of the illegal aliens, transporting them to their destinations in vans and trucks. Some illegal aliens, however, do not migrate so far and remain in Arizona.

It is not known how many people in Arizona are of Mexican-American descent. The most recent figures indicate almost 19 percent (Figure 4.1) or about 442,000 people, but the true figure is probably considerably higher. The major reasons for the uncertainty are that census takers do not do an efficient job in the poorer income areas and many Mexican Americans avoid being counted since they are illegal aliens.

Many Mexican Americans are still dependent on being farm laborers (Figure 4.15). In the American South, the large plantations were broken into smaller units, and the sharecroppers live on the land, but the situation is different in the Southwest. There, the farm units have always been large, and farm laborers commute from towns and cities to the fields. Laborers are seldom hired by the farmers but by labor contractors, usually Mexican Americans, who act as intermediaries between farmers and laborers (Figure 4.16). Many Mexican Americans could get jobs performing menial tasks in industry, but it is understandable why many of them do not like such work: in industry they are the lowest paid employees, they are assigned to sweatshop conditions or menial tasks (such as sweeping floors), and they are frequently the butt of ethnic jokes. Workers in the fields, however, can usually work as fast or as slow as they wish, since they are commonly payed on the

FIGURE 4.15. Two young illegal aliens at work in a field. They are picking rocks out of a field, placing them in the bucket, then walking to the edge of the field to dump them.

FIGURE 4.16. Bus belonging to a labor contractor.

basis of the amount of work done, and in harvesting some crops, such as onions, they work close enough together that working is almost a social occasion. Field workers are also paid daily, which is greatly appreciated.

Today most of the Mexican Americans live in cities and towns. They are mostly unskilled workers in an urban area or work in fields beyond the city limits, and they usually make up the lowest income group in the city. They do not live in slums around the inner core but in large communities, called barrios, scattered about the city. The term barrio is often used to indicate a Mexican-American slum area, but the word really means a neighborhood or district (thus in Latin America, a barrio can be a wealthy or a poor section of a city). A barrio in Arizona, although usually a poor area, is very different from slum areas in the East with their tenements and overcrowding. A barrio is a low-density area, and the structures are mostly single-family dwellings. Although many houses are in poor condition, they are owner occupied, and the owners take great pride in their ownership. There is also a great variety of housing, and a few middle-class families have chosen to build rather expensive homes in a barrio rather than to move out and leave their family and friends.

Another type of Mexican-American community is very similar to the barrio, but it is in a rural setting; this is the *colonia*. A *colonia* starts out usually as a small labor camp for migrants, a place where a farmer will let the laborers live while they work in the nearby fields. If the farmer sells lots, some workers build houses and live there permanently; eventually a store, bar, and filling station open; and another *colonia* exists. There are thousands of these *colonias* between the Gulf Coast and California, and there are many in Arizona. Some *colonias* near large cities are eventually engulfed by those cities, and then they are considered barrios rather than *colonias* (Figure 4.17).

Racially, Mexican Americans are mestizos, a racial mixture of mostly Indian and some Spanish. There was always much racial mixing in the Spanish New World. Such mixing in Mexico was fostered by the government and the church, and as early as 1501, Spanish men and women were encouraged to take Indian mates. It was felt that this was an excellent way of allying the Indians to the Spanish cause, as well as to Christianize and civilize them, but many mestizos were born "without benefit of clergy." Members of the first generation almost invariably allied themselves with the Spanish side and

FIGURE 4.17. A typical scene in the rear yard in El Mirage, a Mexican-American community. Unlike the Americans who came from the East, few Mexican Americans try to maintain grass lawns.

considered themselves as Spaniards. There were soon many mestizos, and they became a very important group, both politically and militarily. After many generations, almost all Mexicans, with the exception of the members of a few Indian groups, began to consider themselves as mestizo. Thus, Mexico is very fortunate to have most of its citizens consider themselves as members of one racial group, as opposed to other Latin American countries, such as Colombia, where the people are of Indian, Spanish, or Negro descent, and there is little mixing or contact between the groups.

A few other racial elements have been added to the Mexican population. As already mentioned, many Chinese went to Mexico, but they were a cohesive group and did not intermarry with the Mexicans. A few blacks, mostly men, were also taken to Mexico, since many of the early explorers had black servants with them, for example, Estevanico. Afro-Indian unions were also considered perfectly acceptable by everyone.

Thus, the Mexican American is a mestizo, and the idea that much of California and the Southwest was settled by "old, proud, Spanish families" is a myth that finds its origin in novels and is perpetuated by the movies and television. As an example of the great mixture of peoples on the Spanish frontier, the first settlers of Los Angeles were nine Mexican Indians, eight mulattoes, two blacks, two Spaniards, one mestizo, and Antonio Rodriguez who was of Chinese ancestry. Such a mixture of peoples was the rule, even on exploring expeditions.

The Mexican Americans are different from other Americans racially, but there are marked cultural differences as well. The majority of the Mexican Americans are Roman Catholic, at least nominally, but so is a large percentage of all Americans. The use of the Spanish language, which is very common in most Mexican-American households, is one of the major cultural differences, and their attitude toward education is another one. In the Latin world, a few wealthy people receive an exceptionally good education, but the masses get only a minimal one. Thus, less pride is taken in and less emphasis is placed on education. That attitude, coupled with the fact that many Mexican-American youngsters cannot speak English when they enter Anglo schools, hurts them academically and in later life when they are forced to make their way in a highly competitive English-speaking country. Many Anglos bemoan Mexican Americans for their lack of a work ethic, for putting work off to take a siesta, but the common sight of Mexican Americans working in the fields at a fast pace, in searing heat, and for only a pittance should end such stereotyping. There are other cultural differences, such as music and food (though Mexican food is now considered a delicacy, and there are even fast-food chains selling Mexican food).

El Mirage, a town northwest of Phoenix, is a typical Mexican-American community. The town initially started as a migrant labor camp. During World War II, it was mostly a cotton patch with a few huts for migrant laborers in the low-lying land along the Agua Fria River—one part of the town is still known as Agua Fria. Originally the houses were mostly one-room units that were rented to the migrant laborers, and during the harvest season each held far greater numbers than they had been designed to accommodate. After each harvest season, the community would almost become a ghost town. The ending of the Bracero program was a real "shot in the arm" for the town, and the migrant labor camp began the transition into a thriving town. With the braceros gone, farm wages rose, local laborers were employed for longer periods of time, and all sectors of the economy benefited, including some few services that had developed, such as bars, grocery stores, and filling stations. Another aid to permanency was the development of Sun City, a wealthy retirement community across the usually dry Agua Fria River, since service and construction industries in Sun City have provided employment for many El Mirage residents. The growth of the metropolitan Phoenix area has reached El Mirage, so what started out as a *colonia* should now more correctly be called a barrio, though the town still has its own government.

Residents of El Mirage are still closely associated with agriculture. Up to 70 percent of those working in the town are employed as agricultural workers, and of those, about 90 percent are "stoop laborers" (Figure 4.18), and only 10 percent are employed as equipment operators or as foremen. The remaining 30 percent mostly work as unskilled laborers in the Phoenix metropolitan area, and many of those work in Sun City. Migrant laborers

FIGURE 4.18. The spot in El Mirage where workers, or stoop laborers, meet in hopes of being hired by a labor contractor. It is about noon, and those still waiting probably will not get a job for that day.

still come to town, and during the harvest season, the town will grow from about 3,500 to over 5,000, which puts an extra strain on a town ill equipped to provide services for even its permanent residents. The Phoenix newspapers often mention the problems that result, such as the town's waterworks breaking down with great regularity.

El Mirage is a very poor town. Many of the streets are unpaved, there are no streetlights, trash is found around the town – especially in vacant lots – and some of the housing for the migrant laborers is very substandard. Many of the residents' homes are substandard also, but they are owner built, and the owners take pride in their homes (Figure 4.17). Although it is a Mexican-American community, almost all buildings are of wood-frame construction, and few, if any, are of adobe. There are no medical or dental offices in the town, and to receive such care in a neighboring town is expensive and one must have access to a car as there is no public transportation. The only other options are to go without health care or to go to a health clinic, which is open only infrequently.

Another problem is employment, or rather the lack of it. About 40 percent of the inhabitants of El Mirage are on relief, and many others receive help from various charitable organizations such as St. Vincent de Paul and the Salvation Army. They are not on welfare by choice. Because of mechanization, there are fewer agricultural jobs each year. This can be seen each morning when potential workers meet at one spot, and a labor contractor comes to solicit and transport the laborers needed. By mid-morning, those remaining realize they will just not work that day. One alternative is to become a migrant laborer, but that is a hard life with little chance for improvement, especially for the young. So many of the people remain unemployed, and at the height of the harvest they put even the children to work to earn enough money for hard times. It's a vicious cycle with hard work, low pay, and little chance for escape. Unionization efforts are strong among the farm workers and can lead to better working conditions, but growers will respond with mechanization and fewer jobs.

INDIANS IN ARIZONA TODAY

Indians are the third largest identifiable group in Arizona today, after whites and those of Spanish heritage. There are about 140,000 Indians living on reservations in the state (Figure 4.19), and Indians compose approximately 6 percent of the total population. As mentioned earlier, Indians are basically young and a fast growing group, but with the great migration of outsiders into the state, the Indian population will probably remain about the same in relation to the rest of the population.

There are few jobs or chances for economic advancement on the reservations. As an example, on the Navajo Reservation there is only one job for

FIGURE 4.19

INDIAN RESERVATIONS IN ARIZONA

Reservation	Population on or near Reservation	First Established	Acreage
Ak-Chin	336	1912	21,840
Camp Verde	460	1914	635
Cocopah	465	1917	1,772
Colorado River	1,745	1865	268,691
Fort Apache	7,686	1871	1,664,972
Fort McDowell	348	1903	24,680
Fort Mohave	383	1880	41,884
Fort Yuma	None in Arizona	1884	(in Arizona)1,773
Gila Bend	357	1882	10,337
Gila River	8,600	1859	372,000
Havasupai	326	1880	188,077
Hopi	5,801	1882	1,561,213
Hualapai	912	1883	993,083
Kaibab	110	1907	120,413
Navajo (in Arizona)	83,000	1868	(in Arizona)9,874,899
Papago	10,542	1916	2,773,377
Pasqua Yaqui	2,008	1978	202
Payson	66	1975	85
Salt River	2,950	1879	49,294
San Carlos	5,979	1872	1,827,421
San Xavier	4,587	1874	71,095
Yavapai-Prescott	68	1935	1,409

Source: Bureau of Indian Affairs, Information Profiles on Indian Reservations in Arizona, Nevada, and Utah, 1978, and from various Indian tribes.

every six willing workers. The jobs that do exist mostly involve working for the federal government or for the tribe. The result is that many try to eke out a living in a modification of the traditional way, but they can't because with better health facilities and little concern for birth control, the reservations are now grossly overpopulated for that kind of life-style. Reservation Indians are certainly no longer in balance with nature. Even those Indians who are trying to follow a traditional way of life still need to purchase some goods, but they have little means to do so. Thus, most live at a very low economic level—their average income is estimated as being 75 percent below the national average. One result of the poverty and of little hope for advancement on the reservations is a migration of Indians to the cities.

There were about 23,000 urban Indians in Arizona in 1970, about 19 percent of the total Indian population. But these are conservative figures, and no one knows the exact number of Indians living in the cities in Arizona. First, the Indians were greatly undercounted by the 1970 census, and second, the Indian population is a floating one, moving back and forth between city and reservation. Also, the overall Indian migration to the cities has been rapid since 1970. Many Indians who have moved to the cities in Arizona are from distant areas, including sizable numbers (over 100) of Sioux, Choctaw, Chippewa, Kiowa, Pueblo, and Cherokee and small numbers of Shawnees, Blackfeet, Tlingit, Pawnee, Yakima, Cheyenne, and many others. The largest numbers, however, are Arizona Indians, such as Navajo, Hopi, Papago, and Pima. There are also reservations near some of the cities, and many of those reservation Indians work and spend much of their time in the cities, but they are counted as reservation Indians.

Phoenix is the greatest magnet for Indians, and it has been estimated that there were 16,000 urban Indians in the Phoenix metropolitan areas in 1970. Those Indians moved to Phoenix in order to have a better chance for a job, as a large number of government institutions there employ Indians (such as a major Bureau of Indian Affairs [BIA] area office, the Phoenix Indian School, the federal Indian Health Service, and various Indian tribal offices), and those institutions have affirmative action programs and give preferential treatment to the hiring of Indians. Other major cities with an Indian population in 1970 were Tucson (about 3,600) and Flagstaff (about 3,000). Many of the smaller towns also had an Indian population, such as Winslow (1,130), Holbrook (532), and Ajo (527).

But migrating to the cities does not solve the problems of the Indians. Urban Indians must face the standard challenges of all urban migrants, and their life is made all the more difficult because they are a racial minority and come from a very different cultural environment. The Indian is generally unprepared to compete in an aggressive Anglo world, and thus unemployment is high, and those who are employed usually have low-paying jobs. Many of the Indian migrants also experience a strong culture shock. The result is often a complete breakdown of society and family ties. Many turn to drink, and alcoholism is a major problem among the urban Indians. Since so many problems are experienced by the urban Indians, the question arises as to the responsibility of the federal government. One argument is that those Indians that have left the reservation and have attempted to integrate themselves into American society should be treated like any other citizen. The federal courts, however, have declared that the urban Indians are eligible for benefits given other Indians, and thus they receive some special services, such as hospital care and various social services.

Since the movement of Indians to the cities has not solved the problem, the government has tried to move jobs to the Indians. Industrial parks have been built on reservations, and tax and other incentives have been used to try to lure businesses to the reservations. But these schemes have largely failed. As an example, an industrial park was built on the Gila River Indian Reservation, not in the middle of the reservation to better serve the Indians, but on the very edge, as close to Phoenix as possible. Also, there are complaints that businesses on the reservations pay little or no tax which gives them an unfair advantage over other businesses. Many of the reservations are very isolated and have few raw materials. Labor is another problem, because few Indians are qualified to handle the managerial and skilled jobs and it is difficult to get qualified whites to move to the more isolated reservations to perform this work. Also, most Indians are not accustomed to working a strict eight-hour day, and absenteeism is a problem for employers.

Reservations make up about a quarter (26.6 percent) of the state, and approximately 140,000 Indians, or about 80 percent of Arizona's Indians, live on reservation land, which is set aside for the Indians and held in trust by the federal government, so it cannot be sold (Map 4.8). Although Oklahoma has more Indians, Arizona has far more full-blooded and reservation Indians than any other state. Each tribe has its own elected officials to manage the tribe's affairs.

Traditionally, Indians have viewed land ownership in a communistic way. No one owned land for exclusive use, rather, the ancestral land was owned by the group and was for the good of all. No one could exclude any member of a tribe from using a portion of the land, and certainly no chief had the right to give away or sell a part of it. Such a system of land control should work well on a reservation, but unfortunately, the federal government imposed the European concept of private property upon the Indians, and today some tribal councils have little control over large sections of their reservations. The General Allotment Act, also known as the Dawes Severalty Act, became law in 1887, and it greatly changed the relationship of the Indians to their land.

The act was passed by people who were sincere in believing that if the Indians were awarded title to land, they could better enter American society and become effective citizens, but the act failed to help the Indians. With its passage, individual Indians on some reservations were awarded 40 to 160 acres (16.2 to 64.8 ha.) of land, and the remainder of the reservation land was opened to white settlement without consultation with the affected Indians (one example of the opening of the Indian lands was the great Oklahoma land rushes). There was to be a period of inalienability, during which the land could not be sold, but after that time it was believed that the Indians would be integrated into American society, so they would be given title to their land to do with as they pleased like any other citizen. Fortunately for the Indians in Arizona, the plan was not carried to fruition in that state, because in the states where it was tried the results were disastrous. Whites, who were better able to compete in the society into which the Indians were thrown, soon began acquiring the land, and many Indians were once again wards of the state, but now they were landless. Between 1887 and 1934 American Indians in the United States lost title to over 90 million acres (36.4 million ha.) of land. Although some land was allotted in Arizona, no Indians in that state were allowed to sell their lands, and no lands

ARIZONA INDIAN RESERVATIONS

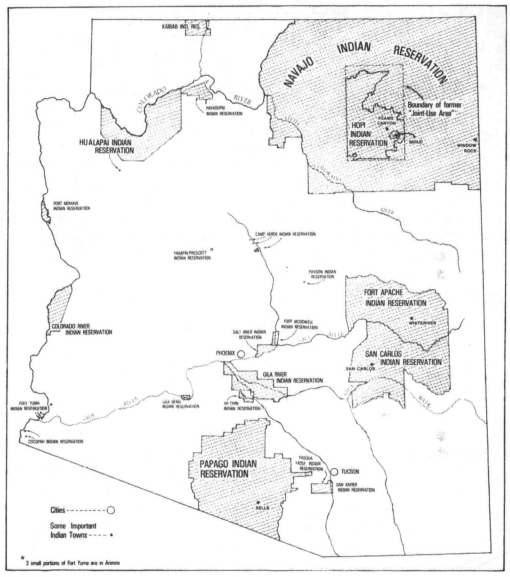

MAP 4.8

were opened to white settlement.

The Indian Reorganization Act of 1934 was an admission that a mistake had been made in 1887, but efforts to get already allotted reservation land that had been acquired by individual Indians back into tribal hands were futile. Fortunately for the Arizona Indians, only portions of the San Xavier,

Gila River, and Salt River reservations had been extensively allotted, as well as small portions of the Navajo and Colorado River reservations. In fact, only 3 percent of the total amount of Indian-owned land in Arizona was allotted under the terms of the Dawes Act. Also, a small amount of individually owned land held in trust was initially acquired under the terms of various homestead acts (such as an outlier of Hualapai land south of the main reservation and the Quechan land near Yuma). Revenue from this private land, much of it leased to whites, goes to individual Indians rather than to the tribe as a whole. Many of the individual allotments have been subdivided through inheritance into very small parcels of land, which compounds the problems of resource management.

Reservations in Arizona today (Map 4.8) can be divided into two broad categories, those for Indians who were basically hunters and gatherers at the time of European contact, and those for Indians who mostly practiced agriculture. The hunting and gathering Indians were the Navajo, Apache, Paiute, Yavapai, and Hualapai, and the agricultural Indians were the Hopi, Havasupai, Pima, Papago, and the various Indian groups living along the lower Colorado River.

The Navajo Reservation is the largest in the United States, occupying 9,874,899 acres (3,999,334 ha., or 15,398 sq. mi. [39,880 sq. km.]) in Arizona, plus an additional 5,268,972 acres (2,133,934 ha.) in New Mexico and Utah. The reservation grew slowly in size, with various sections added at different times, and more land may be added in the future. The total population is now over 153,000 and about 83,000 of the total live in Arizona. The headquarters is at Window Rock, Arizona, very close to the border with New Mexico.

The Navajo were never defeated or subjugated by the Spanish. They were hunters and gatherers when the Spanish arrived, but slowly they accepted certain Spanish and Southwest Indian traits, such as the herding of sheep, silversmithing, the growing of maize, and weaving. The Navajo culture changed slowly and in an orderly way. When the Americans entered the Southwest, the Navajo could effectively resist the white intrusion, and there was trouble between the two groups until Kit Carson was sent to subdue the Navajo in the 1860s. He went through the Navajo territory like Sherman through Georgia, destroying crops, animals, and homes. The last free Navajo retreated to Canyon de Chelly, and when Carson entered that area the end of the Navajo revolt had come. There were about 8,000 Navajo at that time, so in terms of sheer numbers, they were very successful in the ensuing century.

The Navajo have a fairly prosperous economy on their reservation. Much of the early economy was based on sheep, and sheepherding still remains important (Figure 4.20), though cattle are beginning to replace sheep. The Navajo unfortunately learned sheepherding very early, and they picked up many bad habits in this early era. The Navajo considered the number of

FIGURE 4.20. Navajo woman driving sheep down a sand dune in Monument Valley. There are several goats in the herd; the goats constantly move while browsing, thus encouraging the sheep to keep moving. Women own the sheep in this matriarchal society.

sheep owned a determinant of wealth, and they had little concern for quality or for the carrying capacity of the land. The result has been a gross amount of overgrazing by poor-quality sheep, and this situation still exists in many places in spite of great efforts by the U.S. government. There is some forestry, and the Navajo Forest Products Industries produces lumber and particle board.

There is also mineral exploitation: oil and gas were discovered on the reservation in the 1920s, and they have provided a great deal of revenue; uranium was mined in the 1950s and 1960s, and it could again become important; and helium, vanadium, and other mineral resources have also been discovered. But coal is the most important mineral extracted today. The use of coal in Arizona goes back in history to the Anasazi who removed about 100,000 tons (90,700 t.), and a great deal of coal has been mined in the twentieth century for reservation use or for sale in the neighboring towns. The coal is of good quality and is low in sulphur, and a large mining operation is being conducted along the north rim of the Black Mesa, where there is one of the largest deposits ever found. The Peabody Coal Company signed a lease in 1964 for the coal deposits on 24,858 acres (10,067.5 ha.) of the Navajo Reservation, and in 1966 it signed another lease for an additional 40,000 acres (16,200 ha.) within the area jointly shared by the Navajo and the Hopi. There is mining in both areas, and they are currently being strip mined at the rate of about 400 acres (162 ha.) per year. There is, however, a strong commitment to return the land to its original condition, and the Peabody Coal Company is also dedicated to hiring Indians. Approximately 850 of the 1,000 employees are Indians, and all are well paid; the lowest starting wage is presently about $10 per hour, and the average wage for employees is $30,000 per year. These Indian laborers are almost all Navajo. The Hopi have generally, for religious purposes, been against coal mining, and they compose only about 1 percent of the Indian employees. A few Indians from other tribes are also employed.

Handcraft industries are also important among the Navajo. Silversmithing was learned from the Spanish, and the Navajo excel at it. It is an occupation mostly followed by men, and it is handed down from father to son. Like sheep and horses, silver was a symbol of wealth, and it was always worn. In the late 1960s Indian jewelry made out of silver and turquoise became popular, and the Navajo have profited greatly. Rugweaving—always done by women among the Navajo—is another occupation of significance, and it too was learned from the Spanish (Figure 4.21). The Navajo rugs seem expensive, but if the time to make a rug is considered, the woman who made it receives very little for her work.

FIGURE 4.21. Navajo woman weaving a blanket, while her mother on the right cards wool. The blanket on the ground is commercially made.

It must be remembered that unemployment is a real problem on the Navajo Reservation, as it is on most reservations. With a labor force of 66,000, 26,000 are unemployed, an unemployment rate of almost 40 percent. The situation is even worse than these figures indicate, for approximately one-third of those who are employed earn less than $5,000 per year. This situation will quickly worsen, for typical of the Indians in the Southwest, the median age of the Navajo today is 19. Many youngsters will soon enter the job market, but no new industry has been attracted to the reservation lately, and one large facility closed after labor unrest.

One striking feature of the Navajo is their dispersed settlement pattern. They live in one- or two-family units, and all of the units are isolated from each other (Figure 2.5). One reason for the dispersal is the traditional herding of the sheep. The sheep are kept in small corrals near the home at night, and during the day they are herded by the women or children. This method of herding makes dispersed dwellings necessary. It is also common for the Navajo to have two homes as a result of the seasonal movement of the sheep; one at a higher elevation for summer use and one at a lower elevation for use in winter.

The Navajo live in hogans that blend in well with the natural surroundings; in the daytime a traveler might pass by without noticing them, but in the evenings a surprising number of flickering firelights are visible, each indicating a Navajo dwelling. The traditional hogan somewhat resembles a small tepee that has been covered with dirt. This older style is not popular today, and the hogan now used is generally six- or eight-sided, built of logs or railroad crossties, and chinked with mud. The top is a dome-shaped cribbed roof with a smoke hole in the center; the fire is built in the center of the single room, and it does a fair job of keeping out the winter cold. The Navajo often live in brush shelters for a large part of the summer. Hogans remain popular with the older generation and as a ceremonial room. Thus, although many Navajo live in rather typical American-style homes, hogans are often found nearby (Figure 4.22). Even in towns such as Flagstaff hogans are often found in Navajo backyards.

After their arrival in the Southwest, the Athabascan speakers quickly began to divide into groups. The Navajo established themselves in what is now northeastern Arizona, and the rest of the Apache split into six distinct groups scattered in the area between either Arizona and the Great Plains or present-day Texas and New Mexico. The two important groups in Arizona were the Western Apache, centered around the White Mountains and westward, and the Chiricahua Apache in southwestern New Mexico and southeastern Arizona. Each of these groups was divided into smaller bands, each independent of the next. The Americans did not understand this division, and when they established the reservation, they forced many Indians from various bands onto the same reservation, including many who were not even Apache.

FIGURE 4.22. A rather typical Navajo home. On the right is a hogan, and a ramadalike structure is on the left. A pick-up truck, rather than a car, would complete the picture.

There are two large Apache reservations in Arizona, Fort Apache and San Carlos. The Fort Apache Reservation is the home of the White Mountain Apache tribe, and it consists of 7,700 Indians and 1,644,972 acres (666,214 ha.). The San Carlos Apache tribe live on the San Carlos Indian Reservation which has fewer than 6,000 residents on 1,827,421 acres (740,105 ha.). Thus both are large reservations with relatively few people. When the Americans entered the Southwest, the Apache were much feared and they were subdued only after much difficulty and bloodshed. They were a people who had traditionally occupied a great deal of land, and in an effort to keep these feared Indians happy, they were given very large reservations on good land that ranges from forest to low desert. It is unfortunate that friendly and sedentary Indians, such as the Pima, were not so well treated.

The Apache have a matriarchal society, which is common among Indians. When a man marries, he moves to his wife's home, and his allegiance belongs to her family. There was some polygamy, but it was difficult for a man to support several wives, so it was seldom practiced. If there was more than one wife, they would be closely related (they would call each other "sister," and could be, but in the Anglo way of thinking they would usually be cousins). Traditionally it was the woman's job to build the dwelling, which was called a wickiup. It was built of long, slender saplings placed in a circle in the ground and folded over and tied at the top, brush and branches were woven into this framework, and the outside was thatched or covered with bark or deerskin. These structures were easily built and quickly abandoned, and a ramada was often the only structure built in the summertime. The wickiup is now rarely seen, but the ramada still exists and often can be seen next to the wood-frame houses in which almost all the Apache now live.

When the Americans entered the Southwest, the Apache made a great deal of their living by raiding, but eventually they settled down to ranching, and their reservations have some of the best ranching land in the state. Much of it is in grass, and the cattle are moved to the higher elevations in summer and the lower elevations in winter. About 35 to 45 percent of the reservations are covered with juniper and chaparral, and a great deal of money has gone into attempts to eradicate these "weed trees." Their elimination could increase the grazing capacity of the rangeland many times over, and although the Apache are considered pioneers in this area, they have little funds to continue the work. Much effort has also gone into upgrading the herds by introducing quality Hereford bulls, and now Apache cattle are considered to be among the finest in the state.

Since the Apache learned cattle raising very late, they had no bad habits to unlearn and have become some of Arizona's best ranchers. The first ranching on Indian land began with the leasing of rangeland to the Anglos. In this way, money was earned and Apache herds were built. Unfortunately, the Anglos had little concern for the future of the land, and it was greatly overgrazed, but the Apache, having learned a hard lesson, now tightly control the number of cattle grazed. Anglo ranchers have long since left the Apache reservations, and ranching is controlled by the tribes.

The Apache have a highly organized system of ranching. For example, there are more than 20,000 head of cattle on the Fort Apache Reservation, and most of them are privately owned. The only tribally owned cattle are in the southeastern tip, and all profit from those cattle belongs to the tribe. The rest of the reservation is divided into eight tracts of land, and each tract is leased to a livestock cooperative association. The only money the tribe receives for the use of this land is the money paid for the grazing permits. Each association is composed of 37 to 130 members, and each member has his privately owned cattle on the land leased by the association. The cattle have two brands, one indicating the owner and the other, the association. The associations are responsible for fencing and making various improvements on the land leased. Each association elects a three-man board of directors, and one of the three is the president; the presidents of the eight associations form a "general livestock board," which oversees the cattle industry on the reservation. Individuals are hired by the general board and by the individual associations to manage the various operations. Since the associations are cooperatives and the cattle are privately owned, each owner must help at roundups, brandings, and the like, or he must pay someone to take his place. All the associations have grazing land at both high and low elevations, and some of the longest cattle drives in Arizona today are when the cattle on the Fort Apache Reservation are driven from one grazing area to another.

Another important activity on the Apache reservations is lumbering, and 8 million board feet are cut annually on the San Carlos Reservation, and about 60 million on the Fort Apache Reservation. (One board foot is a length of lumber 1 ft. long [30.48 cm.], 1 ft. wide [30.48 cm.], and 1 in. thick [2.54 cm.]). The Fort Apache Timber Company (FATCO) is the richest enterprise on that reservation. It has two mills and a total capacity of 92 million board feet, the total allowable annual cut, since cutting is done on a sustained yield basis. The company buys the standing trees from the tribe, and that payment is the only money directly received by the tribe for their timber, about $2 million annually. The industry employs 260 Apache and a few Anglos, and the payroll is important to the tribe's overall economy. Southwest Forest Industries has a long-term lease with the White Mountain Apache tribe for a mill site at McNary. Timber was initially purchased from the tribe, but once the Apache developed their own forest industry this mill had to purchase timber from national forest land north of the reservation. In late 1979 the mill burned, and it is now abandoned.

A third industry of significance is recreation, and the Apache sell hunting, fishing, and camping permits. However, the tribes must build and maintain the campgrounds, and just hauling off the trash is expensive, so many Indians now question the value of this industry. They resent outsiders on their land, especially since the program brings in little revenue, causes problems, and is expensive to continue. The White Mountain tribe on the Fort Apache Reservation has been fairly receptive to the idea of outsiders using Indian land for recreation, and cabin sites are leased, and Sunrise Park, a large ski resort, has been built. About $3 million of federal money and $1.5 million of tribal funds have gone into building Sunrise Park, but the resort has yet to make a profit, and every year it costs the tribe more and more money. It has great potential, but many problems have yet to be overcome, particularly labor problems.

There are few other sources of income on the Apache reservations. The handcraft industry is not important, only some baskets and beadwork are sold, and agriculture is not significant, since only a few thousand acres are farmed and much of that is in gardens. But the money from outside agencies is very important. On the Fort Apache Reservation, of the $12 million of personal income, $7 million comes from federal and state government jobs and programs.

The other former hunting and gathering Indians in Arizona are not as well off as the Navajo and the Apache. The Hualapai live in a remote portion of the state, and there are no urban areas nearby. Except for livestock raising, there is little economic activity on the reservation, so about 43 percent of the labor force is unemployed. The new freeway, Interstate 40, will bypass the reservation, and that fact will hurt the one filling station and the ice cream shop that now cater to the through travelers.

The Yavapai Indians are sometimes called the Yavapai-Apache because their life-style resembles that of the Apache, but they are very different since they are Yuman speakers as opposed to the Athabascan-speaking Apache. The Yavapai live in central Arizona on several scattered reservations: Campe Verde, Fort McDowell, Yavapai-Prescott, and one of the newest reservations in Arizona, the Payson Indian Reservation established in 1975. There is little or no industry on any of these reservations, but all are near towns or communities where employment can be obtained.

The last of the nonagricultural Indians are the Paiute who live on the Kaibab Reservation on the Utah border. It is a very isolated reservation with few employment opportunities; in 1977 there were 44 Paiute Indians employed and 32 unemployed. There is some grazing, a little agriculture, and some government jobs, but otherwise there are few opportunities for employment.

The traditional agricultural Indians – the Hopi, Havasupai, Pima, Papago, and the various Indian groups living along the lower Colorado River – were generally peaceful Indians who caused few problems for the American settlers, and these Indians were farming only relatively small amounts of land when Arizona was acquired by the United States. Although these Indians did hunt and gather over vast tracts of land, the Americans considered them basically agricultural Indians and therefore entitled to enough land to sustain them agriculturally, but little more. As a result, the agricultural Indians received only small reservations, though it was generally good agricultural land. The Papago were the exception to this rule. They inhabited a desolate corner of Arizona, and although they did practice agriculture, there were very few spots suitable for growing crops, and thus, hunting and gathering remained important in their economy. As a result, the Papago have a large reservation, about the same size as the state of Connecticut.

The Hopi, meaning "the peaceful people," are a large and well-known tribe in northeastern Arizona. In the past they were commonly called Moqui, a term given to them by the Navajo that meant "the dead ones," but the name was greatly disliked by the Hopi. They have a strong claim to the land because they are the descendants of the Anasazi, the ancient occupants of the land. But unfortunately for the Hopi, the Navajo – a large, dynamic, and aggressive tribe – moved into northeastern Arizona just prior to the American period and displaced the Hopi from much of their ancestral land. There have always been bad feelings between the two tribes, and even today there is still conflict over the use of the land. The Hopi must share their land with the Navajo, and one of the villages on Hopi land is inhabited by Tewa Indians from the Pueblo country of New Mexico. These Indians were invited to live in this village, called Tewa, about 1700, in exchange for helping the Hopi fight surrounding enemy tribes. The Tewa still live in the village, and they have retained their cultural identity.

The Hopi are a conservative group and cling to the old ways. The Hopi still practice their old religion, and the kivas, kachinas, and dances have real meaning to them—they do not exist for the sake of the tourists. Many Spanish padres worked among the Hopi but with virtually no success. Seventy-three Hopi in one village were baptized after the conquest of New Mexico by the Spanish under Diego de Vargas, though those baptized continued to practice the old religion in conjunction with Christianity. The conversion upset other Hopi who killed the Christians and destroyed the village. It was a black day in the history of the usually peaceful Hopi. Mormons were to later work among the Hopi, and often various Protestant groups, but all had very limited success.

A great deal of the Hopi religion revolves around rain. Rainfall in this dry and barren region is both very important and very unpredictable, so the concern of the Hopi for rain is understandable. One of the first Spaniards to visit them misunderstood this concern and thought they worshiped water. Much of the Hopi religion also revolves around nature and the relationship of humans to the environment. It is not a formal religion with a strict dogma, and it is very difficult for a Hopi to try to explain it to an outsider. Important in this religion are the kachinas, who are supernatural beings living on top of sacred mountains, such as the San Francisco Peaks. Men will dress like kachinas and impersonate them at dances, and kachina dolls are made and given to the children so that they learn to recognize the various kachinas. Some kachinas (known as chief kachinas) are unchanging, and others are created as new contacts are made. As an example, one is supposed to represent Estevanico, and others could easily represent other more familiar characters, such as Colonel Sanders of fried chicken fame. About 200 kachinas are known, and as some develop, others are forgotten. The kachina dances are held only during the first half of the year, from January to mid-July, and masked impersonators do not take part in the ceremonies held in the last part of the year (such as the snake dance). The kachina dances play an important role in Hopi life. Young boys are initiated into groups, or societies, that perform the dances; each of these groups is important within the social framework of the tribe, and each has its own kiva. Girls are also initiated, but their role is only to aid the kachina activities; a female kachina is always impersonated by a man.

Many of the Hopi insist on living in traditional dwellings (Figure 4.23), and a walk into some of the old communities—such as Oraibi, the oldest continuously occupied town north of Mexico—is like walking into an Indian village several hundred years ago. As a result of this desire, the U.S. government considers 672 of the 882 Hopi dwelling units to be substandard. Many other Hopi, particularly the young, want to enjoy the amenities of life, so some more-modern houses have been built and some towns, such as New Oraibi, have running water and electricity. Most of the Hopi houses are

FIGURE 4.23. Hopi houses on the very edge of a high mesa.

built of stone, the one building material that is plentiful. In comparison to the ruins of some of the prehistoric dwellings, the Hopi houses are poorly built. The ceilings are constructed in the traditional way: poles are laid over juniper beams and then covered with brush, and clay is put on top of all. The inner walls are also plastered with clay. Formerly, entrance to the lowest story was by means of a ladder from a hole in the roof, but now all the houses have doorways leading to the outside. The Hopi men provide the building material, but traditionally it is the woman's job to build the house. By so doing, she owns the house, and should she deem her husband unworthy, she would put his saddle, blanket, and other possessions outside, and his only recourse would be to return to his family.

The Hopi live in compact villages (pueblos). This cultural trait can be traced back to the Anasazi and is the complete opposite of the Navajo settlement pattern. Until the Pueblo revolt of 1680 most of the Hopi towns were near the base of mesas, but then the Hopi moved their towns to the mesa tops to better defend themselves against Spanish reprisals. The need for these protective sites has passed, but their locations have become so interwoven with tradition and religious ceremony that the Hopi cling to them, in spite of the lack of sewage facilities, running water, and other amenities.

The Hopi villages are built on bare rock, immediately on the edges of vertical cliffs. The villages are on the ends of mesas that project southwestward from the Black Mesa, and these mesas are called—with reference to their location from the BIA agency to the east in Keams Canyon—First, Second, and Oraibi (or Third) mesas. Each village was built in an irregular fashion and evolved around a court. Rooms were added by building on the roofs of the first dwellings, and the upper stories were terraced back from the court, to a height of four stories at the most. Other courts sometimes developed if the town grew. It is in these courts where the various ceremonial dances take place.

Agriculture is the way the Hopi survive in their harsh environment, and at the present time there are about 12,000 acres (4,860 ha.) under cultivation. It is a desolate area that would repel white farmers, but one in which the Hopi thrive. Most of the field work is still done by hand, and the agricultural techniques have changed little since Anasazi times. The Spanish did introduce fruit trees, such as apple, apricot, and especially peach, and these trees have become important in Hopi agriculture. They are planted in the sandy belts that border the mesas, especially on the northern and eastern slopes to avoid the prevailing southwesterly winds.

All the traditional crops, such as corn, beans, melons, squash, and pumpkins, remain important. Many other field crops have been introduced, such as wheat, watermelon, and various garden crops, and they have had a major impact on the diet of the Hopi. The Hopi cornfields look quite different from the typical American cornfields, because the corn is not in rows but rather planted in bunches about 6 ft. (1.8 m.) apart. Planting is done by digging a hole – now generally by using an iron pipe – and the seeds are planted quite deep (for example, 8 in. [20.3 cm.] or more below the surface for corn) in order to have more moisture for germination. A handful of seeds (8 to 10) is planted in each hole (Figure 3.3) in order to allow for losses caused by mice, birds, cutworms, and wind-driven sand. Another reason for planting in bunches is that the plants protect each other: the plants on the exposed windward side will be wind whipped and produce little, but those in the center will do well. The corn planted is a drought resistant strain that grows only a few feet tall. To help protect the crops, reeds and rocks are set as windbreaks to ward off the drifting sand, though today a punched-out tin can often serves the same purpose. In spite of all the precautions and elaborate agricultural techniques, the Hopi have learned through hard experience that crops cannot be depended upon, so they keep enough corn in storage to tide them over a year or two.

The fields are widely scattered and are owned by the various matrilineal clans. One type of field is a small garden-like terrace below a spring. (The springs and seeps are caused when water falls on the mesa, enters a porous sandstone cap, and then is channeled to the surface along the edges of the mesas.) Other fields are watered by flood irrigation, and they are located mostly in washes or near the mouths of washes, and sometimes there is a small earthen dam to hold the water or to deflect it to a field. Fields are also commonly found in sandy areas, because moisture is conserved below the sandy layer. This type of field is shifted as the wind blows the sand from the fields to adjacent areas. There are also some fields that are irrigated by modern well methods.

The Hopi are not completely dependent on agriculture because they do have other sources of income. The Hopi have increased their herds of sheep and cattle, but unlike the Navajo, the Hopi do not greatly overgraze the land because they have a deep respect for the environment and want to be in balance with nature. There are a few other animals of economic significance, particularly horses, donkeys, and hogs. Pottery and basketry were once important handcrafts, but now they are secondary to silversmithing. The Hopi styles of silverwork are highly desired and very expensive. Quite different from Navajo jewelry, the Hopi jewelry is generally overlay work, and turquoise is not used. This unique jewelry style developed in the 1930s. The selling of kachina dolls to tourists is another significant handcraft industry.

There are mineral resources on the Hopi Reservation, and the tribe has received well over $3 million over the years in oil leases, although no oil has yet been found. Coal leases generate about $1 million per year, and there is interest in uranium exploration. It must be remembered, however, that the Hopi are a very religious people and there is a strong bond between them and nature. As a result, many of the Hopi are violently opposed to digging into the "mother earth"; to them it is wrong to mine such wealth. The Peabody Coal Company has a policy of hiring Indians at their large mine on the north edge of the Black Mesa, but only a handful of Hopi work there, and the Navajo, who do not have the same strong religious bond to the land, dominate the work force.

The Hopi and the Navajo have unfortunately been embroiled in a bitter dispute over land. Washington officials were ignorant of many aspects of the Indian situation in Arizona, such as the bad feelings between the Hopi and their traditional enemy the Navajo, and made a major blunder when setting aside land for the Hopi. The Executive Order of December 16, 1882, was designed to give the Hopi a reservation, but the wording of the order establishing the Hopi Reservation included the statement that it was to be used by "such other Indians as the Secretary of the Interior may see fit to settle thereon." That phrase was to come back to haunt the Hopi, and it is the cause of problems that exist today. The Hopi have lived where they are now found for at least 2,000 years, and the Navajo are recent invaders, entering the area in about A.D. 1500 (about the time the Spanish arrived). The Navajo aggressively spread out, and as more and more land was added to the Navajo Reservation, they came to encircle the Hopi. The solution to the problem, a reservation exclusively for the Hopi, was not accomplished, and "other Indians" (Navajo) encroached on Hopi land – and did so legally. Just two years after the establishment of the Hopi Reservation, the resident agent reported that Navajo herds were overrunning outlying Hopi fields and that in these contacts the Navajo were "almost invariably the aggressors." The Navajo simply filtered across the reservation boundary, and they were never challenged by the Hopi or by the federal government. Eventually, Navajo stock were allowed to range over all of the Hopi land except for Land

Management District No. 6 (Map 4.9). That land management district became the only area reserved for the exclusive use of the Hopi, and the rest of the Hopi Reservation was designated as "joint-use " land.

The joint-use land was thus for the use of both Navajo and Hopi. However, it was used almost exclusively by the Navajo, and this fact understandably upset the Hopi, especially since the Navajo constantly overgraze and, to the Hopi, seem to have little concern for the environment. The Hopi have made various attempts to claim their share of the joint-use area since 1960, but they have always been repulsed by the more numerous, aggressive, and politically powerful Navajo. A bill was passed by Congress in 1974 to divide the joint-use land. The two tribes negotiated for six months with no result, so a federal mediator drew the boundary line (Map 4.8). The location of the line causes problems, particularly to the Navajo. About 40 Hopi will be forced to relocate to their side of the line, but up to 4,800 Navajo will be displaced. The government will greatly help in the relocation, paying a large cash amount or a house to encourage a family to move. Many

NAVAJO-HOPI JOINT-USE AREA

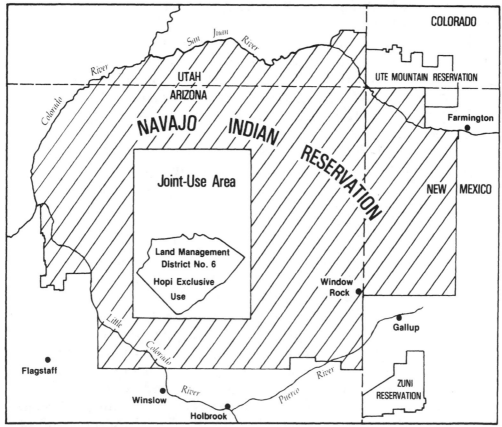

MAP 4.9

of the Navajo, however, are traditionalists, and they have little use for money because they live in a barter economy. These Navajo will resist all efforts to move them, and violence is a real possibility. Another problem is that there is no place for the Navajo to go to continue a herding economy, since all of the Navajo Reservation has been overgrazed.

Another small but very interesting agricultural Indian group in northern Arizona are the Havasupai. They live at the bottom of a deep side canyon of the Grand Canyon, in probably the most beautiful spot in a beautiful state. Their reservation is also the most inaccessible Indian reservation in the United States. There is a road 60 mi. (96.5 km.) long (much of it dirt) leading northward from Route 66 a few miles east of Peach Springs, but at the road's end there is an arduous hike of 8 mi. (12.9 km.), or horseback ride, down to the Indian village. Most tourists entering the canyon continue another 3 mi. (4.8 km.) to a campground deeper in the canyon.

The reason the Havasupai live at the bottom of this canyon is there is both water and level land. Havasu Canyon, also known as Cataract Canyon, is a deep canyon cut by Havasu Creek whose waters enter the Colorado River. About 10 mi. (16.1 km.) from the mouth of the canyon, several large springs (totaling 28,200 gal. [106,737 l.] per minute) emerge from the canyon floor, and from that point the stream flows permanently. Water from this stream is used to irrigate the fields of the Havasupai, which are located just below the springs on level land about half a mile (.8 km.) wide and 3 mi. (4.8 km.) long.

The ancestors of the Havasupai were Yuman speakers who moved onto the Coconino Plateau from the west about A.D. 600. Soon after 900, they began to inhabit the bottom of this canyon in the summertime, practicing agriculture. Winters were not spent in the bottom of the canyon for a variety of reasons that included lack of fuel, scarcity of wild food, short winter days, and high humidity caused by the stream. Thus, the winters were spent on the plateau, hunting and gathering wild food. The total population has been approximately 200 for most of their history, but now about 325 live on the reservation.

The first contact the Havasupai had with the Europeans occurred when Father Garces visited them in 1776. By that time, however, they had already been indirectly influenced by the Spanish, for in the Havasupai settlement Garces noticed cattle (some with mission brands), European-type fruit trees, and red cloth – all had probably been acquired from the Hopi. The first extensive Havasupai contact with Americans came in 1879 when lead and silver were found below the Indian village. Many Americans came into the area, but although scars from the mining activity are still very visible, little wealth was realized. Because of this mining activity, and because ranchers interfered with the Havasupai on the plateau, the government granted them a reservation in 1880. It was only a small reservation, and two years later its size was further reduced, and all of the land was at the bottom of the canyon. The Havasupai continued to use the plateau, however, especially in the wintertime, and in 1974 their reservation was extended to give them 185,000 acres (74,925 ha.) of plateau land, though much of the planning and use of the land is controlled by the federal government.

Most of the homes in Havasu Canyon are small frame houses. There are about 70 dwellings, and 50 of them were flown in by National Guard helicopters in 1977. The traditional home of the Havasupai, called *hawas,* was a lean-to covered with twigs and branches and open on one side. They were often destroyed by flash floods, but they were quickly and easily replaced. For winter use, some very substantial homes were built. Shaped like cones, they were built around a framework of forked poles and were partially underground, and some modified versions of this style still survive. An attempt to improve housing after a severe flood in 1910 was a failure. The government built small frame houses to replace the *hawas,* but the Indians refused to live in them because of poor insulation, and soon all but three were used for storage. A generator on the plateau supplies power, which comes down to the village through 3 mi. (4.8 km.) of cable, and a sewer system has been built. The use of electric refrigerators and freezers has replaced the ancient practice of storing food in caves in the canyon wall, but the acculturation of these people is far from complete.

Agriculture remains important to the Havasupai. The traditional crops of corn, beans, and squash remain important, and many introduced crops, particularly fruit such as apricot, peach, and plum, are significant in their diet. There is one tractor in the settlement, but it usually sits idle as it is often broken. Traditional ways remain, such as the Corn Planting Ceremony, which is performed before a family plants the corn with hoes.

Unlike the Hopi, Havasupai land ownership is passed through the male line of a family. If a man dies and leaves several sons, they share equally in the land. But if he has no sons, it goes to his brothers and their sons, not to his daughters. His widow can have usufruct of the land (she can continue to live on and use the land, without harming it), but only with permission of the new owners, and if she marries, this privilege is lost.

An important source of income comes from tourists traveling through the reservation to visit three waterfalls below the settlement. A small fee is charged each tourist, and the money helps the tribe and contributes to trail maintenance. Money is also earned by renting horses and mules, which are used for riding and as pack animals to bring in the tourists and their supplies. This is a lucrative business, although it has some drawbacks. Many Indians resent the tourists walking around taking pictures, and in order to feed the horses a considerable portion of the agricultural land must now produce fodder crops, such as alfalfa, rather than food crops. As a result, the Havasupai must now purchase a great deal of their food in the outside world.

The waterfalls that tourists come to visit are the Navajo, Havasu (Figure 4.24), and Mooney falls. Each is quite different, and all three are quite

FIGURE 4.24. Havasu Falls. Note the travertine terraces.

beautiful. The highest is Mooney Falls, which has a vertical drop of 200 ft. (61 m.). The water is blue-green and very clear. Travertine, a form of calcium carbonate, builds up near the tops of all three falls, making curtain-like lips through and over which the water cascades. Below the falls are deep pools formed by the travertine building reeflike dams. Any logs, rocks, or brush that gets lodged in the stream are quickly covered with calcium carbonate, and soon bluish travertine dams develop to form more pools. The vegetation is very lush; cottonwoods and other large trees grow on the valley floor, and moisture-loving plants such as ferns grow near the waterfalls. The vertical cliff walls form a striking backdrop, and beyond them there are other barren cliffs.

Changes are coming to the Havasupai. The population on the reservation is now over 325, and the increase has caused some economic problems. There are few jobs down in the canyon, and many of the Havasupai will have to go to the cities and towns for employment. The only hope for increasing the reservation's income seems to be in the tourist industry; 10,850 tourists visited the reservation in 1976, and the saturation point seems to be about 15,000. Selling handmade baskets used to be important, but today basketry is a dying art. Education has also brought many changes. The village school now goes through the sixth grade, but children in the higher grades must go to distant boarding schools. It is to be hoped, for the sake of the Indians, concerned outsiders, and tourists, that Havasu Canyon can remain a unique and beautiful place to live and visit.

A completely different group of agricultural Indians are the Yaqui, a group of Mexican Indians from southern Sonora. Relations between the Yaqui and the Mexicans were particularly bloody, with many massacres on both sides. Violence first flared up in the 1730s when the Mexicans encroached on potentially rich Yaqui agricultural land. The violence reached a peak, and was almost constant, throughout the late 1800s. In the first decade of the twentieth century, the Mexican government attempted to settle the conflict by simply arresting all the Yaqui and selling them to owners of henequen plantations in Yucatán and sugar plantations in southern Mexico. This action caused many Yaqui to flee to the United States, but the Yaqui in Mexico continued to revolt until as late as 1927.

The Yaqui that fled to the United States settled in four communities, three near Tucson and one, named Guadalupe, southwest of Tempe. Since the Yaqui are not historically U.S. Indians, the U.S. government originally felt no responsibility for them, and they were not given reservations. But a bill passed on September 18, 1978, does give them recognition, and now they possess the Pasqua Yaqui Reservation near Tucson. Most now earn their living as farm laborers or are employed as unskilled laborers in the towns near the reservation.

The Pima Indians are agricultural Indians of central Arizona, and they are descendants of the Hohokam. Linguistically they are closely related to the

Papago, though the Pima (the "river people") have a tradition of being true agriculturists, and the Papago (the "desert people") are considered as having only a little agriculture and more of a roving way of life. The Sobaipuri, a former Pima group along the San Pedro, retreated from that area because of the presence of the Apache and were absorbed by the Pima living along more western streams. Living among the Pima today are a group of Yuman speakers, the Cocomaricopa, who moved from the lower Colorado River area, because of constant harassment by their neighbors, and settled peacefully among the Pima. This move happened sometime in the 1700s, and today these people, known as the Maricopa, still retain much of their cultural distinctiveness.

The Gila River Reservation was established in 1859 for the Pima and Maricopa Indians. Soon after the Civil War, however, white farmers around Florence began damming the Gila and diverting the water to their fields. Soon the ancient and dependable canals downstream went dry, and the peaceful Indian farmers had to fight poverty and starvation. In the 1870s about 1,200 Pima and Maricopa Indians were invited by the whites to move north and settle along the Salt River. The whites wanted these Indians as allies in fighting the Apache, and there was plenty of water and land along the Salt River. The whites soon realized that the land was potentially very valuable and tried to oust the Indians, but it was too late, and the Salt River Reservation was established in 1879. Since the Pima were peaceful, agricultural Indians who only farmed a small area, they were not given particularly large reservations. They did hunt and gather over a wide area, but the Americans did not think that they really "used" the extra territory and therefore thought that they should be content with a smaller amount of land. At the time of the Spanish contact, probably 40 to 50 percent of the food eaten by the Pima was hunted and gathered in the desert at some distance from their fields, and the loss of that land certainly affected their economy and standard of living.

The Pima attitude toward wealth was very different from that of the whites, and when the first Americans began to arrive, the Pima willingly shared their food, particularly their wheat and corn. The Pima were subsistence farmers who saw no reason to accumulate wealth or goods; accumulation only meant that one had to work harder. It was also a Pima custom never to refuse a request for aid, so they were very helpful to early white travelers and settlers.

Agriculture has always been important to the Pima. The traditional crops of the Pima were maize, beans, and pumpkins. Several crops were successfully introduced by the Spanish, but the crop that made the greatest impact was wheat. Wheat was introduced in the Tucson area by Father Kino in 1694, and by 1770 it rivaled maize in importance all along the Gila River. Wheat did not supplant maize, but the records show that it was planted in the fall and harvested in the spring so it complemented the growing of summer maize. Because wheat made this double-cropping possible, it was a particularly valuable addition to the economy of the Pima.

Prior to the arrival of the whites, there was a dependable flow of water in the Gila River, but since their arrival there has been a lack of water, which has posed a hardship on the Pima. To help with this situation, Congress authorized the building of Coolidge Dam and stated that the dam was "for the purpose of providing water for irrigation of land allotted to the Pima Indians on the Gila River Reservation, Arizona, now without any adequate supply of water." This authorization was challenged in court, and the resulting Gila River decree, which is still operative, means the Indians must share the water that is stored by Coolidge Dam. At present, 80,000 of the 372,000 acres (32,400 of 150,660 ha.) that make up the Gila River Reservation are irrigated, and about 16,000 of the Salt River Reservation's 49,294 acres (6,480 of 19,964 ha.). As is the case in much of Arizona, if more water were available, much more land could be farmed. There are excellent groundwater reserves, and some of those reserves are used in agriculture, but groundwater is a resource the Indians are trying not to abuse.

The land that was allotted to individuals is a special problem. Some of the allotted land is tied up in heirship proceedings, and some of it has been splintered into parcels too tiny for farming. So the owners and heirs lease the land to others (mainly whites) who have the money to assemble enough land and equipment to farm it profitably. It would be better if the tribe had control over all the allotted land and could operate it as one large, efficient farm for the benefit of all the tribal members.

The Ak-Chin Reservation is a smaller one for both Pima and Papago Indians. Their main occupation is farming, and the chief employer is Ak-Chin Farm, a tribal enterprise. About 6,000 acres (2,430 ha.) are under cultivation, and although much of the rest of the reservation is suitable for farming, water is the limiting factor. Since there is no major stream on the Ak-Chin Reservation, the Indians are dependent on groundwater. However, the fact that the water table is falling is causing great concern, and the tribe has been forced to cultivate less land. When the Indians began farming in 1962 the water table was 200 ft. (61 m.) deep, and now it is closer to 600 ft., (182.9 m.), and of the 38 wells, 15 have had to be abandoned because of a lack of water. The whole position of agriculture in their economy is now being questioned by the tribal members. They are using their water at a rapid rate, with no chance for natural recharge, and they are earning only a little on their agricultural crops in exchange. Farming does provide employment, but the wages are very low—the only well-paid employees are the whites who hold the administrative positions. Many feel that benefits are not worth the costs, though water from the Central Arizona Project could change the situation.

The Gila River Reservation and the Salt River Reservation are in excellent positions with regard to transportation, population centers, and areas of economic activity. Unlike many of the isolated reservations, they are on the fringes of metropolitan areas, with all the prospects (and problems) that this contact entails. Several large industrial parks have been built on the reservations, and others are in the planning stage. Many small industrial, recreational, and educational areas have been leased to the whites, and with federal aid the Indians are establishing various businesses. Many job opportunities are also available in the neighboring towns and cities.

It must be remembered, however, that in spite of these advantages, there is still unemployment and poverty on these reservations (Figure 4.25). On the Gila River Reservation in 1978 there was a labor force of 2,442, but about 50 percent of that labor force was unemployed; and of those employed, 75 percent earned less than $5,000 per year. As on many other reservations, there is an overdependence on government jobs; on the Gila River Reservation over 50 percent are so employed. The annual average family income also remains low; it is $4,800 on the Gila River Reservation (5.5 persons per family), and it remains far below the national average, as on all reservations.

The Indians along the southwestern border of the state, such as the

FIGURE 4.25. A typical older, rather poor home on the Gila River Reservation. The tamarisk trees in the background require very little water, and they are commonly found in Indian and Mexican-American communities (see also Figure 4.17).

Cocopah, Quechan, Mohave, and Chemehuevi, were basically agricultural Indians when the first whites arrived, especially those Indians in the south. The Spanish attempted missionary activity in the area, primarily in an effort to develop land communication between California and Sonora, but they were soon repulsed. Their first strong contact with the outside world occurred when the Mexican forty-niners (often called Sonoras) abandoned the goldfields of California and drifted toward the lower Colorado River for placer mining. Not long after this, these Indians began losing their cultural identity, especially the Quechan who quickly moved into a wage economy.

The Colorado River Indian Reservation is different from the other reservations because it contains Indians from various tribes. It was originally set aside primarily for the Mohave, but not exclusively because it is also for any of the many other Indians of the Colorado River drainage who may desire agricultural land. After World War II, other Indians, principally Navajo and Hopi, began establishing themselves on the Colorado River Reservation as farmers, primarily raising alfalfa. Soon the Mohave began to fear that they would be outnumbered, so they began to oppose further colonization by outsiders, and it was stopped in 1957. The Hopi living on this reservation have been isolated from their homeland and are now greatly acculturated. That some of their Hopi brothers have lost the old ways and their cultural identity is something the conservative Hopi in northeastern Arizona find difficult to accept.

The Indians along the California border are among the more impoverished, but there is much potential for improvement. The land is rich, the sun almost always shines, and since the construction of Hoover Dam, there has been a dependable water supply and the threat of flooding has ended. Thus, the agricultural potential is very attractive, and not surprisingly, agriculture is important. However, much of the farming is done on land leased to whites, and much of the irrigable land lies idle. With federal help, the Indians are now trying to increase the agricultural acreage and to have the farming done by the Indians. They should succeed. Another great potential source of income for most of these Indians is the development of recreational sites along the Colorado River and its lakes. Some of these sites have been developed, and others are in the planning stage. Both directly and indirectly, the sites should bring in money to the tribes.

The Papago Indians are on three reservations, San Xavier, Gila Bend, and Papago. The San Xavier Reservation has often been characterized as an "Indian suburb," because it is close to Tucson, the Indians there are greatly acculturated, and most are employed in the city as unskilled laborers. The Gila Bend Reservation has few such advantages. The U.S. Army Corps of Engineers has a flowage easement on 90 percent of the land, and no permanent structures can be built on that land. There is little agriculture (about 700 acres [283.5 ha.]) and no industry. Compounding the economic problem is the fact that there are few jobs in Gila Bend, the nearest town.

The Papago Reservation, the second-largest reservation in the United States, is located in an isolated section of Arizona. To the south is Mexico, and to the west are the large and rather barren Luke-Williams Air Force Range and the Organ Pipe Cactus National Monument. Topographically, the reservation is in the heart of the Basin and Range district, and it contains broad desert valleys and narrow abrupt mountains, with only the Baboquivari Mountains (7,730 ft. [2,356.1 m.]) on the eastern edge reaching any great height.

Although descended from the Hohokam, the Papago were desert dwellers when the Spanish arrived. They practiced only a little floodwater farming, and about 75 percent of their diet came from hunting and gathering. Their settlements were seasonally abandoned as they moved from their fields to sites where wild food could be gathered. Since they were migratory, the Spanish missionaries made little impact on them. They received their Christianity from the Indians to the south, and they blended certain features of the new religion with their old to form a new one called Sonoran Catholic. By no stretch of the imagination could it be called Roman Catholic.

When the Spanish arrived in the Southwest, each Papago village was very widely spread out in a seemingly haphazard way, with trails going into the desert and ending at family units. Each of the village units was autonomous, and several of them made up a dialect group. Eleven such dialect groups existed in Kino's time, and essentially they still exist. Each unit was independent, was made up of related Indians in a patrilineal society, and had an accepted leader. There were no chiefs of the dialect units, much less a chief of all the Papago.

The typical house at that time was the *ki* (pl., *ki ki*). The *ki* had a long history that can be traced, with changes, back to the Hohokam period. It was built by the men in a small pit with four central posts to support the roof, and all around the inside of the pit small pliable poles were placed vertically in the ground. These poles were folded over at the top in beehive fashion and lashed to the four central posts, and the whole structure was covered with brush, straw, and arrowweed. Earth was then heaped on top. A firepit was near the center, but apparently there were no smoke holes. The Papago spent most of their time out-of-doors and only used the *ki* in time of bad weather or for storage. The *ki* survived into the twentieth century, and the last one was built in 1930.

Each family had not only a *ki* but several other structures as well. For shade, a ramada was built of four large posts to support a flat roof of small sticks, arrowweed, and the like, with a covering of earth over the top. Storehouses have also been found, and they were rectangular structures similar to the ramada, except that walls were added. The walls were made of vertical sticks, such as ocotillo trunks or cactus ribs, and sometimes plastered with mud. Agricultural and gathered products were stored inside in large baskets and pottery jars. The *vato*, or cooking enclosure, was the

FIGURE 4.26. A rather typical family compound on the Papago Reservation. A fence surrounds several buildings, identifying the family area, and there are few, if any, trees and hardly any vegetation.

final family structure. It was a windbreak of sticks, such as ocotillo, placed vertically in the ground in a semicircle around the open hearth. Within this area the pots, jugs, and other cooking utensils were also kept.

Since Spanish contact, the housing of the Papago has slowly changed. The *ki* is no longer built, nor is the *vato*. Cooking is still commonly done outside, but now it is done under the ramada or next to it. The oldest house style found on the Papago Reservation, and far and away the most common, is a rectangular one with the door centered on one of the long sides and windows on the short sides and often flanking the entryway (Figure 4.26). The Papago build such houses with wattle and daub (now considered the "traditional" home), adobe, and "Pima sandwich"—a newer construction technique.

The wattle-and-daub house is particularly common in the southwestern part of the reservation, the most conservative area. This type of house greatly resembles the old storehouse and probably evolved from it. It is crude wattle-and-daub construction and can't compare with the excellent wattle and daub found in southern Sonora. The adobe houses are very similar in size and shape, and they are mostly found on the eastern edge of the reservation, the most progressive area. Adobe brick was introduced to the Papago when the San Xavier Mission was first built. They eagerly learned the technique of making and using adobe, and it is still a common building material. Pima sandwich construction is a relatively new building tech-

FIGURE 4.27. An older home of Pima sandwich construction on the Gila River Reservation. This one uses railroad crossties to support a traditional, heavy roof, and the wood holding the mud is just sticks gathered from the desert. Note the bed beneath the clothesline to the right; much of the summers are spent out-of-doors (see also Figure 4.25).

nique, and it probably first evolved among the Pima further north in the 1920s. In Pima sandwich construction, large posts, preferably railroad crossties, are placed vertically in the ground several feet apart. Small boards (1″ × 4″) or branches are nailed to the outside and inside edges of the posts to form a skeleton framework, and thick mud and straw are placed within this framework (Figure 4.27). A Pima sandwich house is much easier to build than an adobe house, and if it is plastered over, it is difficult to tell one from the other at a distance. Unfortunately, Pima sandwich is not as thick as adobe and does not insulate as well. Usually this type of house has a traditional, heavy roof, so the vertical posts have to be large to support the weight. If an American-style frame roof, covered with sheet metal or asphalt shingles, is to be built, a lighter framework, perhaps using 2″ × 6″ boards, is sometimes used. Pima sandwich construction is now particularly common along the northern edge of the reservation, and it is gaining acceptance everywhere. This construction technique requires only a little money and an exchange of labor within families. Some newer Pima sandwich houses are built on concrete slabs and have plumbing and electricity and other modern features such as aluminum window frames (Figure 4.28). The result is a good, solid house, the building of which costs little but helps strengthen traditional family ties.

Modern housing is gaining some acceptance on the reservation. The federal government, through the efforts of the Papago Housing Authority

FIGURE 4.28. A new and modern Pima sandwich home on the Gila River Reservation. It is built on a concrete slab with 1" x 4" lumber purchased from a lumber yard, and it has interior plumbing, aluminum window frames, and an American-type gabled roof. This type of house, built by the owners, perhaps should be encouraged.

and the Housing and Urban Development Agency, selected several growth centers and has built Anglo-style houses at those sites. The Bureau of Indian Affairs and the Public Health Service also offer advisory and technical assistance. However, even though the government helps with the costs, these houses are still too expensive to build, too difficult and complicated for the unacculturated Papago to maintain, and too expensive to heat and cool. Some Papago also must live right next to Papago who are not family members, and few of the Papago like the Anglo style of house. These and other problems have resulted in few successes in improving the Indians' houses. A few mobile homes, owned by some wealthier and more acculturated Papago, are now also used.

The Indians on the Papago Reservation have traditionally maintained strong contacts with the Indians and Mexicans to the south, and until the 1950s, the Papago were more "Mexicanized" than they were "Americanized." After the Gadsden Purchase, Americans moved onto land that is now part of the reservation and established mining camps and ranches, but apparently these white settlements had little impact on the Papago. The Papago were finally given a large reservation in 1916, but until World War II, they had few contacts with the Americans. Since that time, acculturation has grown stronger and stronger.

Many problems now face the Papago, unemployment being among the worst. Many of the Papago are forced to go to neighboring towns and cities such as Tucson, Ajo, and Casa Grande for employment. Many others are employed at very low wages as stoop laborers in fields off the reservation. There is some traditional farming on the reservation, and with federal help, the Papago are beginning a large farming project in the southwestern portion of the reservation. This is the Papago Farms effort, and they plan to get 20,000 acres (8,100 ha.) into production by using groundwater and to employ many Papago. Ranching is important in the economy, but overgrazing is a serious problem.

The Indians of Arizona are scattered across the state, with the greatest numbers on reservations, and Indian populations have risen greatly in the last few years. Only a few of the Indians can earn a living in anything resembling the traditional way, but there are only a few jobs on the reservations for the others. The result is unemployment and gross underemployment, and this situation leads many Indians to migrate to the cities. There they are unprepared to cope and compete, which results in a breaking down of their society and in serious alcohol problems on and off the reservations.

The Indians of Arizona are not a cohesive group. They speak many languages, come from very different backgrounds, and have different cultures. Some are farmers, others are not; some are conservative and resist change, others are dynamic and progressive and quickly adapt to modern ways; and many tribes are traditional enemies of one another. In this setting, it is easy to see why there had been little cooperation among the Indians as a whole.

MINOR ETHNIC GROUPS

The minor ethnic and racial groups in the state include only a few people. Approximately 96.3 percent of the population is of white, Spanish, or Indian heritage, which means there are only a few people in other categories. Blacks and Orientals are important to the state and are very visible; within the white category there are also some distinctive minor ethnic groups such as Jews and Basques.

The blacks in Arizona make up about 3 percent of the state's population, there are about 70,000 of them, but they are not a powerful group since they have little political influence or economic control. Although many came to the state as agricultural workers, they did not own any land, nor did they control any of it in a sharecropping system. Thus, unlike the Mormons who also arrived as agriculturists, the blacks have not been responsible for specific changes in the landscape. However, the blacks do make up a distinctive cultural and racial group, and they have a unique history.

The history of the blacks in Arizona is an old one, and it certainly goes back to the arrival of the first whites. Estevanico, the black slave, was in the Arizona area with the earliest European explorers—either with Cabeza de

Vaca or, if that expedition did not enter Arizona, when he guided the expedition led by Marcos de Niza. Later, many blacks also went to Arizona as cowboys. Many plantation owners moved to Texas to become ranchers when the cotton lands in the East deteriorated, and they took their black slaves with them. After the Civil War, many of the blacks continued as ranch hands and took part in the great cattle drives. They spread throughout the West with the cattle industry, and certainly many entered Arizona. One, known as Nigger Jim, went from Texas to southeastern Arizona with John Slaughter's herd of cattle. He was a giant of a man, who was greatly respected in the area round Tombstone and is particularly remembered for having gotten into a boxing ring with John L. Sullivan (he was promptly knocked out after landing the first blow). Another, known as Nigger Jeff, is remembered because he was involved in conflicts between Texas cattlemen and New Mexican sheepherders, and especially for one shoot-out in St. Johns. These two men stood out because they were unusual, and because their names included an offensive epithet, but the vast majority of the cowboys who rode through the West, both black and white, were ordinary men who are now nameless. The only cowboys who are remembered today are those who stood out because they were particularly honorable and respected, or especially because they were violent and criminally inclined.

Another fairly sizable group of blacks to enter Arizona in the early years were in the military. Many blacks served in the armed forces during the Civil War, and in 1866 an act was passed that allowed blacks to serve in the peacetime army. Six black regiments were authorized, four infantry and the 9th and 10th Cavalry, and three years later the four black infantry regiments were consolidated into the 24th and 25th Infantry. All four black regiments entered Arizona, mostly after the Indian fighting had ended, but they had distinguished themselves in Indian fighting in many places before their arrival, particularly on the Great Plains from the Canadian border all the way into Mexico. These regiments served mostly on the frontier, and they had an excellent record in spite of a lack of decent housing, food, or arms. These black troopers were known as buffalo soldiers, a name apparently given to them by the Indians because their coloring and hair reminded the Indians of the buffalo. The name was not derogatory and was accepted by the black soldiers.

Many people who moved to Arizona in the early American period came from the American South. The movement westward was natural, and it included blacks as well as whites. In the census of 1870 there were 26 blacks in the territory, but the number quickly grew, and by 1900 there were 1,848. In the earliest era they engaged in various skills and professions, but that soon changed, and by 1890, 94.7 percent of the blacks who were employed worked as domestic and personal servants. The black males left that occupation in the early 1900s, but as late as 1920, 87 percent of the black women who were employed still worked as domestics. Only a few are still employed in such jobs today.

Industry and agriculture became important areas of employment for the blacks in the twentieth century. About the time of World War I, people were needed to pick the cotton in Arizona, and, of course, those who knew best how to pick cotton lived in the South. Fliers were distributed widely in the South, particularly in Texas, Oklahoma, Louisiana, and Arkansas, encouraging people who knew how to pick cotton to move to Arizona (Figure 4.29), and many blacks, as well as whites, moved to Arizona. By the time of World War II, however, the blacks were moving away from stoop labor to employment in industry. By the 1970s the transition was essentially complete, and the blacks left working in the fields to the Mexican Americans.

The blacks in Arizona have traditionally been urban dwellers. At least three out of every four blacks live in either the Phoenix or Tucson metropolitan areas, and the rest live in other towns and cities across the state. In the early cotton era, all of the towns in south-central Arizona, such as Casa Grande and Eloy, had large numbers of blacks who would commute to the fields. Mexican Americans and poor whites from Oklahoma and Texas picked cotton as well, and on Saturday nights, after everyone had been paid and had had a few drinks, violence in those towns was more the rule than the exception. In Eloy in the late 1940s, for example, seventeen people were killed in a twelve-month period. The blacks began leaving the cotton industry in the 1950s when cotton picking machines were introduced, and the blacks soon began leaving the smaller towns for the better economic opportunities in the larger cities. Today most blacks in Arizona live in the cities and have the lower-paying industrial jobs. Their educational levels are low, and unemployment, especially among the young, is very high.

There are very few people of Chinese descent in Arizona, approximately 5,000, but they are quite influential. The Chinese began filtering into Arizona in the 1880s to work on the railroads or at various jobs in the mining towns. Most came to Arizona by way of California, though many came via Mexico where they had migrated earlier. Whereas many thousands of whites have migrated to Arizona in the twentieth century, few Chinese have moved into the state in this century and they have small families, so their numbers have changed little. Thus, they were more visible, represented a larger percentage of the population, and were more significant in the prestatehood days than they are now. Unfortunately, little is known of the Chinese and the early problems they faced, because they were a shy, stoic, and quiet people who tried to avoid publicity.

Very few of the Chinese in Arizona were ever associated with "coolie labor" in the fields, and they have not been farm owners. There were apparently cultural reasons for this. Most of the Chinese who migrated overseas left urban areas in southern China, especially Hong Kong and Canton. These people were mostly merchants, and wherever they settled, be it

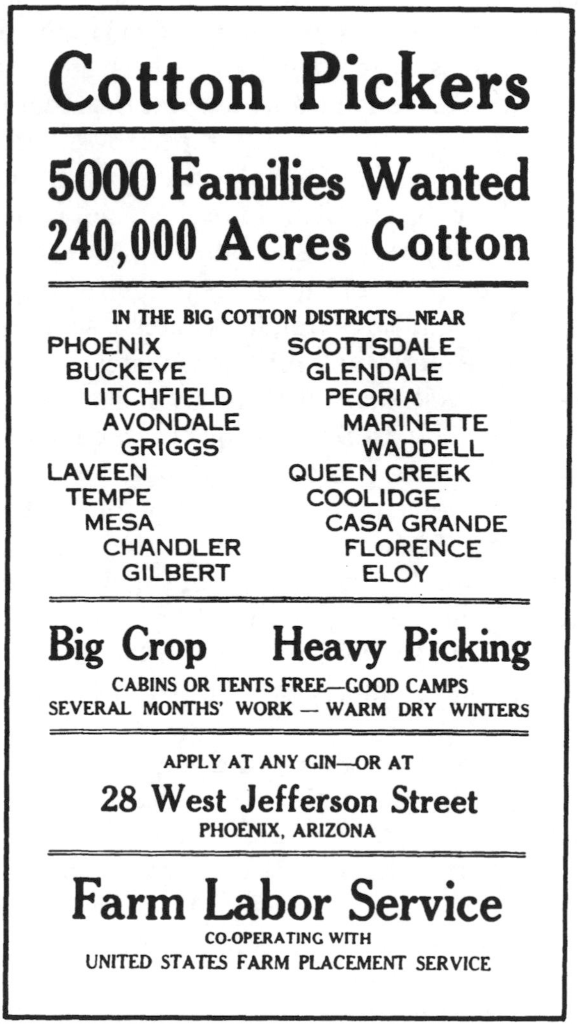

Cotton Pickers

5000 Families Wanted
240,000 Acres Cotton

IN THE BIG COTTON DISTRICTS—NEAR

PHOENIX	SCOTTSDALE
BUCKEYE	GLENDALE
LITCHFIELD	PEORIA
AVONDALE	MARINETTE
GRIGGS	WADDELL
LAVEEN	QUEEN CREEK
TEMPE	COOLIDGE
MESA	CASA GRANDE
CHANDLER	FLORENCE
GILBERT	ELOY

Big Crop   Heavy Picking

CABINS OR TENTS FREE—GOOD CAMPS
SEVERAL MONTHS' WORK — WARM DRY WINTERS

APPLY AT ANY GIN—OR AT

28 West Jefferson Street
PHOENIX, ARIZONA

Farm Labor Service
CO-OPERATING WITH
UNITED STATES FARM PLACEMENT SERVICE

FIGURE 4.29. The type of flier posted throughout much of the South encouraging cotton pickers to move to Arizona. Source: Malcolm Brown and Orin Cassmore, *Migratory Cotton Pickers in Arizona* (Washington, D.C.: Government Printing Office, 1939), p. 72.

Singapore, Manila, Saigon, San Francisco, or Phoenix, they became, with time, small merchants. When the Chinese first arrived in the Southwest, they worked in the towns or in the various railroad or mining camps as cooks or in laundries. They soon became small, independent businessmen by opening restaurants, laundries, and grocery stores in the towns and cities, and they eventually became fairly prosperous. Today many are still in these businesses, and some have become big businessmen, such as wholesalers of fresh produce.

The Chinese encountered a great deal of discrimination and resentment from their neighbors, and the other merchants resented their success. In the 1890s in Tucson, the Chinese merchants were forced to move to the southwest part of the city. In Phoenix, they concentrated more and more in south Phoenix, serving the blacks and the Mexican Americans. The Chinese are a closely knit group, and their aloofness from the people they served compounded the problem. Although living in the community, they were not a part of it; they married among themselves, did not attend the local churches, and were not members of the social organizations. The local residents resented the Chinese and felt they were being used by these foreigners, who lived in the backs of their stores and ate produce that was spoiling or could not be sold. Consequently, violence was not unusual, and hardly a year went by without one or two Chinese being shot in a holdup. The Chinese restaurant owners solved the problem by consciously leaving the central area of a city and moving further out so that they were less visible. However, because of competition from the supermarkets, the grocery store owners were forced to remain as outsiders in the black and Mexican-American neighborhoods.

The Chinese in Arizona are now engaging in other occupations. They still own many small grocery stores, but the number of those stores is declining as the young people do not enter that business. The educational levels among the Chinese are very high, and many of the young people are now entering the professions. The small grocery store, operated by a Chinese man and his wife with the aid of their children, will be a thing of the past within the next generation. However, their cooperativeness and close family, business, and religious ties will make the Chinese a distinct cultural and racial group for a long time into the future.

Another Oriental group is the Japanese Americans although they are few in number, approximately 2,000. The migration of Japanese to the United States was slow to develop, mostly because of a stay-at-home policy of the Japanese government. As late as 1884 the Japanese consul in San Francisco reported only 80 Japanese in California, and most of those were students. The Japanese began to arrive in substantial numbers—over 1,000 per year—in the 1890s, and they arrived with a threefold handicap because of place, race, and time. There had been a great deal of anti-Oriental feeling in California for a long time, and with the passage of the Chinese Exclusion Act in 1882, limits had been put on Chinese immigration. There was a lack of jobs at that time, and the Orientals were blamed for it. All this occurred when the Japanese began migrating to the United States, so they were not welcomed. After 1900, anywhere from 4,300 to 10,000 Japanese were migrating to the United States per year, but in 1908 a gentleman's agreement between the United States and Japan placed the responsibility for limiting this migration on the Japanese government, and the number of those migrating quickly dropped.

Many of the Japanese immigrants became involved in agriculture, though some who entered northern Arizona in the late 1880s were employed in lumbering and mining. Like the Chinese, the Japanese valued education and were hard working, and they could be described as having the "Protestant work ethic" even though they were not Christians. They soon began buying small parcels of land and became intensive agriculturists, noted for their ability to produce large amounts on a small amount of land.

That tradition still survives in Arizona. There are several small Japanese farms that produce vegetables, and they are particularly noted for their flower gardens, several of which are in the Phoenix area (Figure 4.30). Flower production in Arizona began in the 1930s, and the shipping of

FIGURE 4.30. A view of one of the Japanese small, labor-intensive farms south of Phoenix.

flowers out-of-state began right after World War II. There were soon five Japanese families raising flowers, and in the springtime their farms, along Baseline Road in southeast Phoenix, became major tourists attractions. These farmers produce mostly cut flowers for sale locally or for shipment to the large eastern cities but all of these farms are now experiencing difficulties. Many of the young people are not entering the business; labor is getting expensive and scarce; and foreign competition, especially from growers in Latin America where costs are less, is hurting the American flower business. These and other problems will probably mean an end to this business in Arizona within a few years, particularly as the younger, educated Japanese Americans enter other professions.

The only time there were many Japanese in the state was during World War II. With the bombing of Pearl Harbor, many Americans on the West Coast worried about a Japanese invasion and what role the Japanese Americans might play if an invasion occurred. Because of this emotionalism and racism, Pres. Franklin D. Roosevelt signed Executive Order 9066 that led to the internment of about 120,000 persons of Japanese descent. An evacuation zone was established, which essentially encompassed the western half of California, Oregon, and Washington and the southern one-third of Arizona (Map 4.10). The evacuation line went along Grand Avenue in Phoenix, and all Japanese Americans south and west of that line were uprooted and sent to relocation camps; many Japanese Americans lost all of their possessions as a result. The relocation camps were not intended to be internment camps but way stations for those willing to resettle in other parts of the country, or temporary homes for those who wanted to eventually return to the West Coast. Although the United States was also at war with Germany and Italy, Americans of German and Italian descent were not put into relocation camps. The Japanese were rounded up and shipped to the camps not because they were disloyal, but because they were of Japanese descent and "might" be disloyal. The record, however, does not indicate a single act of disloyalty or sabotage on the part of any Japanese American during the entire war, and many volunteered for military duty and fought in the Italian campaign.

The War Relocation Authority established 10 camps, 2 of which were in Arizona. One camp was on the Gila River Reservation and was named "Rivers" after Jim Rivers, the first Pima Indian killed in World War I. That camp was divided into two sections. The other camp was on the Colorado River Indian Reservation, and it was named "Poston" after Charles Poston, the "father of Arizona." Poston was divided into three sections. Interestingly, BIA officials organized these relocation camps. Both relocation camps were in very hot and arid portions of the state and had no air-conditioning, and the residents (prisoners?) at Poston nicknamed the three sections there "Roaston, Toaston, and Duston," which probably reflected conditions there. The Poston camp held approximately 17,800 persons, and Rivers held about

13,300, which at the time made these two camps the third and fourth largest cities in the state, exceeded in population only by Phoenix and Tucson.

The Japanese made the best of a bad situation at the camps. Both camps had rich soil and an availability of water, and hogs, poultry, and especially vegetables were produced at each. The high alkali content of the soil at Poston initially posed problems, but those problems were solved by leaching. Almost all work at the camps was done by the residents, and they governed their own affairs. In some ways, the camps soon came to resemble towns because there were schools, business services (such as watch repairing, tailoring, and so forth), athletic events, and churches. Their homes, however, did not resemble those of towns because everyone lived in barracks, and everything was surrounded by a fence.

JAPANESE RELOCATION CAMPS IN ARIZONA

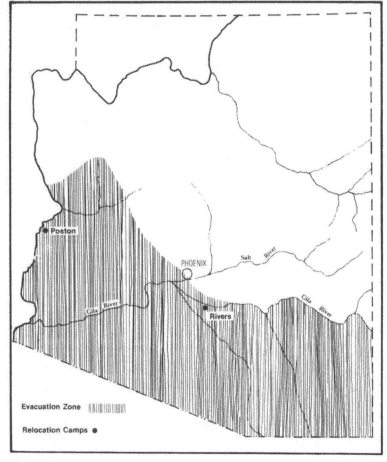

MAP 4.10

The war was going very much in favor of the United States by 1944, and the fear of the Japanese Americans had greatly waned. Many residents of the relocation camps had already left for work and relocation elsewhere, and there was little further need left for the camps. Between October 1 and December 15, 1945, all the relocation camps in Arizona were closed, and most of their residents returned to the West Coast. Little is left of this attempt by the Americans to confine a group because of their ethnic background. All of the buildings at Rivers were sold to the local Indians for a dollar each in 1964 and were moved elsewhere. Only the concrete slabs remain. At Poston, however, two of the three camps still exist, and they are used for various purposes. Some of the buildings are still occupied, but now the residents are Indians.

Jews were another ethnic group that entered the area very early. Jews have traditionally been only a minor part of Arizona's population, less than 1 percent, but their importance far outweighs their number. Most of the Jews who came to Arizona had at least some money, and through their hard work, business experience, education, and intelligence, they quickly established themselves as businessmen. Jews on the frontier were not usually the cattlemen, farmers, or miners but rather the store owners, bankers, and entrepreneurs who lived in the fledgling towns and cities. Almost every town in Arizona had at least one or two Jewish merchants, and they played an important role in bringing the various ethnic and racial groups on the southwestern frontier together.

The first Jew known to enter the Arizona area was Herman Ehrenberg. He was an adventurer who left his native Germany, fought for Texas in its battle for independence, went to California, and in 1854 drifted into Arizona. He was instrumental in one of the first mining booms in Arizona after the American aquisition, at Tubac. He was typical of the Jewish people who went to Arizona in this early era in that he came from California, because many Jewish merchants in California sought greater opportunity in Arizona. At that time steamboats plied the Colorado River (Figure 3.9), and the Jewish merchants used the steamboats to import goods. La Paz, now a ghost town, was the first settlement along the Colorado River, but Ehrenberg later became the major commercial center on the river, and it was laid out by another early Jewish merchant, Michael Goldwater. By the later 1860s, Jewish merchants from the East began to enter the territory, and William Zeckendorf was the most important member of this early eastern element. The two groups met in Tucson; generally speaking, Jews west of there had a strong California connection, and those east of Tucson had contacts with eastern merchants.

Most of the early Jews in Arizona were store owners, but some were in other occupations. Many entered the banking, hotel, or restaurant businesses, and one man, Michael Wurmser, amassed over 6,000 acres (2,430 ha.) of land, much of it farmed by Mexican Americans. Several, like

Ehrenberg, were mining promoters. The Clifton-Morenci area was largely developed through the activities of two Jewish families, the Fruedenthals and the Lesinskys. They developed the copper mines in that area, and in the process built Arizona's first smelter and railroad.

Social and political activities were very important to the early Jews in Arizona. They were prominent people, and they joined many social and civic organizations. They were not only well known, but they were also greatly respected for their business integrity and superior education by all the peoples on the frontier, and, thus, they made good political candidates. As concerned citizens, many Jews accepted the political challenge, and they have been important in politics in Arizona ever since.

Religion was apparently of little consequence to the early Jews in Arizona. This attitude was not unusual, and it was probably the case with many of the Americans during this era. Persons of the Jewish faith were always encouraged to marry other Jews, and it is traditional in the Jewish faith that it is the mothers who determine the Jewishness of the children. There were, however, very few marriageable Jewish women on the frontier, so many Jewish men married Mexican and Anglo girls, and although their children carried Jewish names, they generally lost the religious heritage. Throughout the 1880s and 1890s Jewish people established various benevolent societies, and gathered on various holy days, but the first Jewish congregation was not established until 1903. The first synagogue in the state was built in Tucson in 1910, and the second was built in Phoenix in 1922.

Jews have come to Arizona, like other Americans, in ever increasing numbers. Whereas the earliest Jews in Arizona had their roots in Germany, many of the later immigrants trace their descent back to eastern Europe. The East European ghetto Jews began arriving in the United States in large numbers after 1881, immediately after the assassination of Czar Alexander II and the resulting pogroms, and those Jews continued to arrive until the Johnson-Lodge immigration bill cut off the flow in 1924. Considerable numbers of these East European Jews have been entering Arizona since World War II. These later arrivals, along with the pioneering Jews, have contributed greatly to the development and growth of Arizona.

The mining towns were noted for their ethnic diversity. In addition to the ordinary diversity of people in the West, there were European miners, and two groups stood out. The first group was made up of Cornishmen (called "Cousin Jacks") and Welshmen who came from areas of Great Britain that had a mining tradition, and they succeeded well in Arizona, usually rising above the position of a common miner. Slavic miners made up the second unique group, and they began arriving in Arizona in the last decade of the nineteenth century. They were a diverse lot, including Poles, Bohemians, Russians, Croatians, and others—particularly Serbians. Bisbee once had over 300 Serbian families, and they built a Serbian Orthodox Church there. Slavs, particularly the southern Slavs, came from impoverished areas of

Europe, and the environment in the Arizona mining and smelting towns was much less forbidding to them than it appears to us in retrospect. Each mining town generally had ghettos where the members of the ethnic minorities lived, but many of the younger generation have since migrated to the state's urban centers, and the ghettos no longer exist.

The Basque make up another interesting ethnic group. Although only a few have settled in Arizona, their impact on the sheep industry has been significant. Few Americans know how to handle sheep, and practically no American wants to spend the summer months herding sheep through lonely mountain valleys, not seeing any other human beings. Sheepherding is a very lonely occupation, but the Basque have eagerly engaged in it, and they have long been identified with it. There were Basque on many of the Spanish exploring expeditions in the New World, so some were in the New World very early, but the Basque sheepherders began arriving in the West about the middle of the nineteenth century. They had excellent reputations as sheepherders and were much sought after by the sheep ranchers. The greatest number of Basque arrived in the United States between 1920 and 1950, and in 1950 the U.S. government enacted special legislation that allows Basque herders to be brought into the country. Most of the Basque came with the intention of staying only a few years and then returning to Spain with a large bankroll. However, many have acquired large herds of sheep and have become permanent citizens. They generally do not own any land in Arizona, only the sheep, and they continued the habit of bringing over other Basque herders. They have acquired grazing rights in the mountains, and in the winter months they keep their sheep in rented pastures on irrigated, lowland farms. Today, few Basque are coming to the West to herd sheep, and they are being replaced by Peruvians and others from South America who have a tradition of handling sheep.

Besides the ethnic groups mentioned, there are many other ethnic groups that have settled in Arizona and have left their mark there. The members of some groups such as the Armenians and the Lebanese, are identifiable and not fully acculturated. People of many other extractions—Irish, German, French, etc.—have come as part of the mainstream of Americans who are flooding the state. A few still meet and belong to organizations that try to maintain the traditional ties, but acculturation is a powerful force, and more and more of these ethnics are joining the mainstream of American life, losing their ethnic identity, and becoming simply Arizonans.

5

URBANIZATION
AND URBAN GROWTH

Arizona is an urban state. Most outsiders think that it is populated by cowboys on the range and that there is an Indian on every street corner, but that is far from reality. Those impressions may have been somewhat true in earlier years, for in 1900 only 16 percent of the population was urban, but today over 80 percent of the people live in cities. The population is concentrated in the large metropolitan areas and in Phoenix and Tucson in particular (Map 5.1). Accentuating this population concentration is the fact that 55 percent of the state's population lives in Maricopa County, and 21 percent in Pima, and most of the other 24 percent live in the smaller towns scattered across the state.

Among the towns in Arizona, Tucson had the earliest start. There were two Indian villages in the area of Tucson when the Spanish entered the area. Father Kino made them *visitas* of the San Xavier Mission, and one of them, San Cosme del Tucson, went through several name changes but, in the end, gave Tucson its name. The actual meaning of the word "Tucson" is open to argument. It is undoubtedly of Piman origin and means either "place of the black spring" or "base of the black mountain." Probably a few non-Indians drifted into the area to take advantage of the irrigation possibilities, but the founding of Tucson dates back to 1776 when a presidio was constructed on the site. About 100 colonists also settled among the peaceful Indians who lived there at that time, and a town was established. Throughout the early period Tucson was a small, isolated frontier town, surrounded by an adobe wall, and economically and defensively dependent upon the presence of the military (Figure 4.13).

Until recent years, Tucson was a slow-growing town. During the Spanish

IMPORTANT CITIES AND TOWNS OF ARIZONA

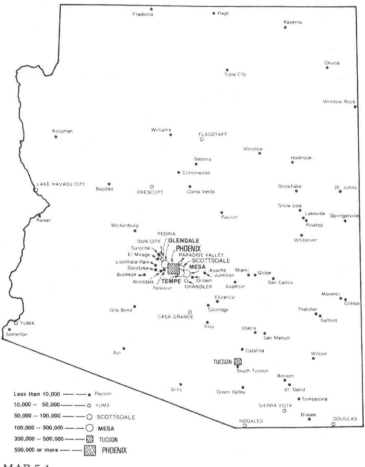

Less than 10,000	———————	•	Payson
10,000 -- 50,000	———————	○	YUMA
50,000 -- 100,000	———————	○	SCOTTSDALE
100,000 -- 300,000	———————	◯	MESA
300,000 -- 500,000	———————	▨	TUCSON
500,000 or more	———————	▨	PHOENIX

MAP 5.1

period the population peaked at approximately 1,000, but because of Apache attacks, the population declined during the Mexican era, and by the time of the Gadsden Purchase in 1853, Tucson had only about 300 residents. Even after the American aquisition, the population growth was slow, and it was not until after the Civil War, when the American government could assert some authority and the railroad was extended to the town, that Tucson began to prosper and grow. The citizens of Tucson were very instrumental in having Arizona made a territory and in separating it from New Mexico, and they thought the capital should be Tucson, the largest and most dynamic city. Most Arizona citizens, however, always felt that Tucson had too "Mexican" a character and was politically too democratic to be the capital, so Prescott became the first capital. In 1867 the capital was moved to Tucson for a ten-year period, but then it was moved back to Prescott. The establishment of the University of Arizona in 1885 helped to stimulate the growth of Tucson in the early years.

The settlement of Phoenix came very late in relation to Tucson. The first settlement began in 1868 3 mi (4.8 km.) east of present-day downtown Phoenix. Abandoned Hohokam irrigation canals were cleared, and an agricultural community evolved to supply foodstuffs to Fort McDowell some 35 mi. (56.3 km.) to the east and to the mining districts in the mountains to the north. Several businesses were started, including a mill to produce flour, and the community was known by several names (Swilling's Mill, Helling's Mill, Mill City, and East Phoenix). The name Phoenix has been used since 1870; inspired by the idea that an ancient Hohokam civilization had existed there and that a new city was rising from the "ashes," the settlement was named after the legendary Phoenix bird.

Several key factors affected the early growth of Phoenix. A year after its founding, it was made the county seat of newly formed Maricopa County, and a certain amount of stability and growth was thus assured. Quite fortuitously the town happened to be in the center of the richest agricultural oasis in the state, and its early growth was closely tied to agriculture and the development of irrigation canals and dams. Tied to the agricultural development was the extension of a rail line into the valley, which allowed agricultural products to be sold on more than just a local basis. First, the Maricopa and Phoenix Railway linked Phoenix with the Southern Pacific at Maricopa Wells about 40 mi. (64.4 km.) to the south. In 1895 the Santa Fe, Prescott, and Phoenix Railway built a line that connected Phoenix with the Santa Fe at Ashfork, and later the Parker Cut-off was built, connecting that line – and Phoenix – to California through Parker, thus avoiding the tortuous route via Prescott and Ashfork. Later a main line of the Southern Pacific was extended through Phoenix (though all through traffic still passes far to the south of the city). With its good rail links to the rest of the territory, and its central location, Phoenix was an excellent site for the capital, and in 1889 the territorial capital was moved there. Agriculture, however, continued as the mainstay of the Salt River Valley economy for the first half of the twentieth century, and Phoenix, which dominated that oasis, prospered.

Phoenix has experienced rather steady growth throughout its existence. In 1870 there were about 300 people living there, by 1880 that figure was up to 1,700 and by 1890, it was up to 5,500. There was continued steady growth through the first half of the twentieth century, and by 1920 Phoenix had begun to outgrow Tucson. During the 1950s there was tremendous growth, partly because of wholesale annexations to the city (Figure 5.1), and by 1978 the population of Phoenix stood at over 690,000 and growing.

Phoenix, unlike Tucson, is largely ringed by large suburban cities, and these cities, like Phoenix, have grown rapidly in the last 40 years (Map 5.2 and Figure 5.2). With the exception of Tempe, all of these cities were

FIGURE 5.1. Air view of the urban sprawl in the Phoenix area as homes are built at a long distance from the city on former agricultural land.

established well after Phoenix was founded. Tempe was founded in 1871 by Charles Trumbull Hayden, who was born in Connecticut and had been a businessman in Tucson since 1858. The beginning of Tempe was along the Salt River where a flour mill was built and a ferry established. The settlement was originally known as Butte City, later Hayden's Ferry, and it was named Tempe in the 1870s by Darrel Duppa because the town site reminded him of the lush Vale of Tempe at the base of Mt. Olympus in Greece. Within 10 years it was a very prosperous little community. In 1885 the Territorial Normal School was established in Tempe, and that school was to grow into Arizona State University. Population growth was slow, and as late as 1950, Tempe had fewer than 8,000 residents. Like all the towns in the Salt River Valley, growth has been spectacular since that date, and the population now stands at over 100,000. In spite of its size, Tempe is basically a bedroom community for Phoenix and a home for the students attending the university.

Two towns in the Salt River Valley were established as religious colonies, Glendale and Mesa. Glendale was established as a Dunkard Colony (Church of the Brethren) by 70 families in 1892. It was planned as the Ideal Temperance Colony of Glendale, and a temperance clause was put in every

URBAN GROWTH OF THE SALT RIVER VALLEY
1940-1978

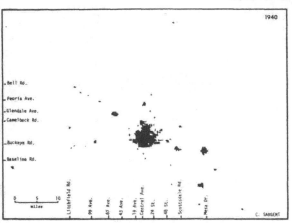

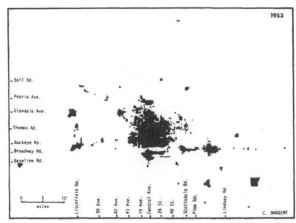

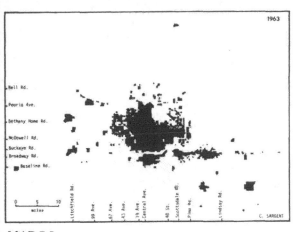

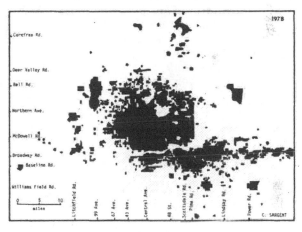

MAP 5.2
Courtesy: Charles S. Sargent, "The Story of the Salt River Valley Urban Growth: Past-Present-Future," 1978.

FIGURE 5.2

RECENT POPULATION GROWTH OF MAJOR TOWNS

IN THE SALT RIVER VALLEY

(Population in Thousands)

	1940	1950	1960	1970	1977	Projected 2000
AVONDALE	x	2.5	6.2	6.6	6.9	36.0
CHANDLER	1.2	3.8	9.5	14.3	22.8	93.0
EL MIRAGE	x	x	n.a.	3.3	4.0	14.0
GILBERT	x	1.1	1.8	2.0	4.0	46.0
GLENDALE	4.9	8.2	16.2	36.2	75.2	155.0
GOODYEAR	x	1.3	1.7	2.1	2.7	36.0
MESA	7.2	17.0	34.0	66.1	115.0	224.0
PARADISE VALLEY	x	x	2.1	7.2	10.2	17.0
PEORIA	x	n.a.	2.6	4.8	11.5	68.0
PHOENIX	65.0	107.0	439.0a	585.0	682.2	1042.0
SCOTTSDALE	1.0	2.0	10.0	67.6	82.0	106.0
TEMPE	2.9	7.7	25.0	64.0	103.0	184.0
TOLLESON	1.7	3.0	3.9	3.9	3.8	19.0
TOTAL MARICOPA COUNTY	186.0	332.0	664.0	968.0	1288.0	2297.0

x - not yet founded or less than 1000

n.a. - not available

a - there was extensive annexation to Phoenix between 1950 and 1960

COURTESY:
Charles Sargent, "The Story of the Salt River Valley Urban Growth: Past-Present-Future," 1978.

deed. In 1895 the Santa Fe completed the line from Phoenix to Prescott through Glendale, and that line later connected with the main line to the north and with Los Angeles, through Parker, to the west. The town gradually grew, became an important agricultural shipping point, and it now dominates the northwest end of the valley.

Mesa was established as a Mormon colony in 1878. A year earlier another Mormon colony had been established in the same general area – called Utahville and later Lehi – but it did not succeed. Both Lehi and Mesa were established as United Order settlements, but that social system did not last. Mesa was platted according to Mormon precepts, and its very wide streets are particularly noticeable. Located on a high tableland above the Salt River, "Mesa" was always the preferred name for the community, but since there was already a Mesaville in Arizona (in Pinal county), the town was named Hayden. The name was later changed to Zenos, after a prophet in the Book of Mormon, and after Mesaville ceased to exist in 1889, the name was again changed to Mesa. It became the center of Mormonism in Arizona after the only Mormon temple in the state was completed there in 1927. Mesa has remained a basically agricultural community, and although relatively small until the 1950s, it came to commercially dominate the eastern side of the valley. Its growth since the 1950s has been spectacular.

Scottsdale is the fourth large suburb of Phoenix. It was founded quite late, in 1894, was originally known as Orangedale, and was occupied by citrus growers and health seekers. The town had a very slow start, and as late as 1950 the population was only 2,000; over 80,000 have since been added. Scottsdale is not just a bedroom community for Phoenix; it has plenty of commerical activity – much of it related to the wealthy elderly – as well as a good deal of light industry.

Many other towns have been established in the Phoenix metropolitan area. Peoria was founded in 1888 as an agricultural colony by settlers from Peoria, Illinois. It came to be on the Santa Fe line between Phoenix and Los Angeles, but since Glendale was the major shipping and commercial site on that edge of the valley, Peoria remained small. Avondale, originally Coldwater Crossing, was established on the Agua Fria River in 1892, and in the 1880s, residents from Ohio settled in the Buckeye district. With the cotton boom during World War I, the Goodyear Rubber Company developed a cotton farm of 12,000 acres (4,860 ha.), and in 1917 founded the town of Litchfield Park to house its workers. It was also common for small towns such as El Mirage and Surprise to slowly evolve as dormitory towns for the large number of laborers needed to work in the surrounding fields. Many such towns began as migrant camps, mostly for Mexican Americans, and then slowly developed into towns, or *colonias.*

Some towns were established as speculative ventures by large landowners. Several individuals managed to acquire large blocks of land within what became the bounds of the Salt River Project, but since federal rules allow an individual to irrigate only 160 acres (64.8 ha.), the land was soon subdivided and "speculative towns" were established. Gilbert, Tolleson, and particularly Chandler are examples of such towns in the Phoenix area. Alexander John Chandler, a veterinary surgeon, came to Arizona in 1887 and with the backing of Detroit businessmen, had acquired 18,000 acres (7,290 ha.) by the early twentieth century. With completion of Roosevelt Dam, Chandler could legally irrigate only 160 acres (64.8 ha.) of his land, so he began subdividing his Chandler Ranch into tracts of 40 acres (16.2 ha.) each. These tracts sold well, many being purchased by California ranchers and fruit growers. He also established the town of Chandler and began selling lots there in 1912, but the town grew only slowly. The San Marcos Hotel, one of the valley's first exclusive hotels, was built here, but although the San

Marcos remained in business until 1979, no other such hotels were built, and Chandler did not become an exclusive resort area. Today the town has a diverse economy, with income from agriculture, industry, and the nearby Williams Air Force Base. The urban sprawl from Phoenix is reaching Chandler, and the town's growth is now booming.

There are many small towns scattered throughout the state. Many of them can be classified according to their origins, and most were founded as mining, agricultural, or railroad towns. Now however, many of them are prosperous communities with well-rounded economies. In many other parts of the country, the railroads went to the towns, but in Arizona and most of the West, the towns went to the railroads. The railroads in Arizona were built to cross the state rather than to serve it, and, besides, there were few towns in Arizona when the first railroads were built. The only major town initially served was Tucson, simply because it was on a logical railroad route. Phoenix and Prescott were completely bypassed at first. In Arizona, as in most of the West, towns were established at strategic spots as the rail lines were built, so many of the towns in Arizona date back to the early 1880s. Along the Southern Pacific route, Benson (1880) and Wilcox (1880) were settled as the railroad extended to those points. Gila Bend had an earlier start (1865), but the railroad tracks were some distance from the Gila River, so the town began to shift southward toward the railroad. Yuma was also settled earlier (1854), but it became an important rail center when the tracks were extended to that point. Along the route of the Santa Fe Railroad, Kingman (1883), Flagstaff (1881), Winslow (1881), and Holbrook (1882) were all established soon after the arrival of the rail line, and along this route apparently only Williams predated the railroad, though Williams did not get a post office until 1881. Parker (1908) was founded where the Santa Fe spur line between Phoenix and California crossed the Colorado, although an earlier town of Parker (1865) had existed 4 mi. (6.4 km.) south of the present town. The railroads still run through these towns but are generally not important in their overall economies. These towns are still on the major corridors of east-west movement, but most of that movement is now highway traffic, and many of the towns are now somewhat dependent on automobile travelers (Figure 5.3).

The mining towns in Arizona are quite distinctive. Most of them are in the mountains, and the streets usually follow the terrain instead of being laid out in the typical grid fashion (Figure 5.4). Mining towns have a quite temporary character about them, especially the dwellings. Miners usually see no need to build pretentious and solid homes, for if the mines shut, the miners will be forced to leave suddenly for other mining camps. Thus, the homes are of low quality and are tucked away on the slopes or in side canyons. This arrangement makes it difficult and expensive to get water and sewage lines to the houses, and many are built on slopes and away from roads, which makes life difficult in this age of the automobile. The more

FIGURE 5.3. A street scene along a major highway entering one of the towns along the route of the Santa Fe Railroad in northern Arizona. This scene exists in all of those towns and illustrates the significance of the through highway traffic to the local economies.

substantial buildings are built on the flat lands in the canyon bottoms. These buildings are usually of brick and are occupied by business establishments or used by the mining companies (Figure 5.5). Most mining towns, unlike the railroad towns, are one-industry communities, and they have a hard time existing if and when the mines close, particularly if they are company towns, such as Ajo (not a permanent town until 1911), Bagdad (1883), and Morenci (1864). It is difficult to estimate exactly when many of these mining towns were settled. Many became boom towns soon after the ore was discovered, and others began as mining camps and developed only fitfully. In southeastern Arizona are the mining towns of Tombstone (the silver ore was discovered in 1878, and it was a town of over 10,000 by 1882) and Bisbee (the ore was discovered in 1877, and it was a major town by 1880) and the smelter town of Douglas (1901). Most of the mining towns are in the central mountains of the state, including Hayden (1910), Winkelman (1903), Globe (1880), Miami (1909), Superior (1875), Jerome (1876), and the mining supply town of Wickenburg (1864).

The farming communities are also significant. Florence (1866), Casa Grande (1880), and Coolidge (1926) are in central Arizona, and in other parts of the state are some farming communities that are primarily Mormon. Examples of these towns are Safford (1874) along the upper Gila; the small

FIGURE 5.4. A typical residential scene in a mining community, this one in Bisbee. Note how the houses and roads cling to the hillsides.

towns along the Mogollon Rim, such as Show Low (1876), Snowflake (1878), and Springerville (which began as a trading post in 1875), but they are now mostly strongly oriented toward recreation and lumbering; and Fredonia (1885) along the northern border.

Besides these railroad, mining, and farming towns, there are others that do not fit into easy categories. Prescott (1864) was established as a mining supply center near an army fort, and it acquired government functions, first as the territorial capital and later as the county seat. St. Johns (1874) was initially settled by Hispanos from New Mexico, though Mormons moved in soon after it was established. It has been split between Catholics and Mormons ever since; it has been a county seat since 1879. Sierra Vista was initially called Garden City (1919) because it produced vegetables for nearby Fort Huachuca. It is now a dynamic and an aggressive little city, but it is completely dependent upon the fort. Nogales, which has appeared on maps since 1859, was established as a shipping and commercial center on the international border, and Payson (1882) was founded as a ranching and supply center. There are many other such small cities and towns scattered throughout the state.

By 1920 the pattern of cities and towns was set. The towns of the Salt

River Valley were dynamic and experiencing a slow but steady growth. Tucson was a large city, but Phoenix was surpassing it in size and overwhelming it economically. Tucson was even frustrated in its attempts to dominate southern Arizona because southwestern Arizona is quite barren, and tied to Phoenix, and southeastern Arizona is closely tied to El Paso, a city that has largely dominated the mining industry in the American Southwest. By 1920 almost all the towns in the state had been established, and only a few unique ones were to be founded later. All of the small towns scattered throughout the state had small economic bases, and none could ever hope to compete with Phoenix or Tucson. Only Prescott had once had great expectations, but by 1920 it had given up all hope of dominance in the state; it is isolated in the mountains, off the main avenues of travel, and only local mining, agriculture, and ranching give it an economic base. Prescott had to give up even its regional commercial aspirations in favor of Flagstaff, which has a college (Northern Arizona University) and is located along the main line of the Santa Fe and at an important highway junction.

Only a few towns have been established within recent years. Page (1956) was initially built to house laborers building the Glen Canyon Dam, and the

FIGURE 5.5. Part of the business district of Bisbee. As is typical of mining communities, the flatlands are occupied by schools, businesses, churches, and mining company buildings. Note the indication of former mining activity in the background.

McCulloch Corporation built the planned community of Lake Havasu City (1963) along the Colorado River. Both of these towns are now viable communities. The other new towns that have been established are mostly retirement communities, such as Green Valley (1963), Arizona City (1961), and the very successful Sun City (1959).

Changes in the population and economy began occurring about the middle of the twentieth century as the cities of southern Arizona started to boom. Winter visitors and health seekers had discovered the warm desert climate in the 1920s, but because of the depression and World War II, it was not until the 1950s that people began arriving in great numbers. Air-conditioning and evaporative coolers had long been perfected, but about that time their use began gaining wide acceptance, and they certainly encouraged permanent residency by people who otherwise would never have settled in Arizona. Light industry, particularly electronics, was being established in the urban areas of southern Arizona by 1950, and it created jobs for many residents and even attracted new residents to fill the jobs. Growth only created more growth as the population began to boom. Once the growth began, Phoenix and Tucson could largely support themselves by that growth, because the construction industry, largely fed by California capital, often employed up to 10 percent of the labor force. Almost all of the new growth was based on the forms and patterns of the past; new houses and developments were built inside and outside the existing towns and cities, which began expanding their boundaries by annexing the new areas.

Phoenix and Tucson thus became large cities. In some ways they resemble all other cities in the West, especially in their broad and expansive setting and dependence upon the automobile, but each is a unique city with its own character. Tucson is older and has much more of a Mexican atmosphere. It is closer to the border, has a higher percentage of Mexican Americans, and has always had strong Mexican-American leadership. Newcomers recognize this aura and appreciate it, and as a result many Anglo homes are built, at least on the outside, to resemble Spanish architecture. Water is scarce and expensive, so there are few attempts at planting lawns and trees; the residents tend to live with the desert rather than to fight it. Tucson deserves its title, "Old Pueblo."

The Phoenix metropolitan area, on the other hand, is very different, because Phoenix has the look and atmosphere of the Midwest. It was settled late and does not have any Mexican or Indian roots, and the original inhabitants were mostly from the Midwest and the South, although many of them came by way of California. These people apparently have never felt they were living in a desert, but rather in an oasis that they have tried to shape to resemble the Midwest, and the availability and cheapness of water have allowed these immigrants to largely succeed in that effort. Shaping an area so that it resembles the homeland is a common cultural trait, but their whole attitude is reflected in the fact that many longtime Phoenicians have never truly enjoyed or come to know the desert, they have only crossed it as quickly as possible on the freeways. The downtown areas of the two cities are also quite different. Tucson has a downtown that is compact and dynamic, with some expansion eastward, but the Phoenix downtown has extended northward for miles along Central Avenue. Also the downtown area of Phoenix is affected by the large number of outlying shopping centers. Valley residents today only infrequently and unwillingly visit downtown Phoenix, and the old core area is changing. The downtown area once served many retail functions, but most of the retail businesses are now gone, and financial institutions and company offices have mostly taken their place. Therefore the character of downtown Phoenix resembles the downtowns of most major U.S. cities, and it is now very different from what it was in pre–World War II days.

Major population shifts are occurring today as many Americans are leaving the colder climates for the Sun Belt, and Arizona is receiving more than its share of newcomers. The vast majority of these new Arizonans are urban dwellers, and they tend to live in the Phoenix and Tucson metropolitan areas. These areas have the jobs, which attracts the younger migrants, as well as the hospitals and other amenities that attract the older migrants. The population in and around these cities is, therefore, growing at a dramatic rate; by 1990 there will probably be another million in the Phoenix area and another 200,000 in Tucson. Sizable chunks of farmland and desert will have to be developed to provide housing for the new residents, and the life-style of all urban dwellers, as well as of all Arizonans, will be affected because these newcomers will also seek out recreational sites in all parts of the state.

6

LAND AND WATER

LAND

It was initially difficult for private citizens in the Arizona area to acquire land from the public domain, but acquisition was very important as it affected so many people and was so crucial to the settlement of the land. No one wanted to settle in an area, make improvements on the land, and then find there was no way to legally own that land. Before settlement could begin on a large scale, some provisions had to be made to allow people to own land.

When Arizona was established as a territory in 1863, it contained over 72 million acres (29,160,000 ha.), and all the land was in the public domain except for 64,000 acres (25,920 ha.) that made up a reservation for the Pima-Maricopa Indians, which had been established four years earlier, and some encumbered lands claimed by right of Spanish or Mexican land grant. There were also some squatters and some mining districts at that time.

Indian Lands

The relationship among the Indians, the land, and the federal government is a complex one. Early on, the Americans developed the attitude that all land "used" by Indians belonged to them and could not be settled by whites; that land was "reserved" for the Indians. But what was considered "use" by the Indians varied greatly from the white definition of it; potential agricultural land that was only periodically hunted on was considered unused land by the Americans but quite important "used" land by the Indians. The government won all such arguments. The establishment of reservations supposedly isolated the Indians to areas where there was only a minimum of white influence and defined where whites could or could not settle and acquire land.

219

The first reservation in Arizona was established in 1859 when the Pima-Maricopa lands along the Gila were surveyed, and the Gila River Reservation, later greatly expanded, was set aside. An extension of that reservation along the Salt River was proposed in 1878, and it was to contain an extensive amount of land, including all the land in the potentially rich Salt River Valley eastward to the San Carlos Reservation. A smaller version of this proposed reservation was established in 1879 as the Salt River Reservation. The second, completely separate reservation established was the Colorado River Reservation in 1865. It originally covered 75,000 acres (30,375 ha.) but was later greatly enlarged. The Navajo Reservation was the third one established, in 1868, and it, too, grew with time. More and more reservations were established for the various tribes, and the last one, the Pasqua Yaqui Reservation, was founded in 1978 (Map 4.8). Indian reservations, once established, were often changed or even eliminated. Most of them were greatly enlarged with the passage of time, though the size of the Apache reservations was significantly reduced, primarily as a result of the discovery that the land to the east, south, and west was good mining land, and thus extensive areas were restored to the public domain. Such reductions to the Apache reservation lands ended in 1883.

The Indian lands are very important to Arizona. They are very extensive in size, comprising over 26 percent of the state—approximately 19,869,152 acres (8,047,007 ha.)—(Map 4.8), and they contain many important mineral resources. More importantly, they may control much of the future water supply of the state.

Spanish Land Grants

The Spanish also had means by which land could be acquired by citizens. All the land claimed by Spain was owned by the crown, and the Laws of the Indies, which were established in the New World in the early 1500s, allowed for land grants. The object of land grants was to stimulate settlement, for both agricultural (*hacienda*) and ranching (*rancho*) purposes. Grants could be given to towns, churches, or individuals, but when a land grant was given, the nearest towns and Indians had to be considered. The Indians were not to be hurt by land grants, and their lands were not to be included in grants given to Spaniards. At that time Spain still had a feudal social system, and a large block of land was of little value unless there were peasants (in the New World this meant the Indians) to work the land, so the law was often broken. Under Spanish law, the Indians were considered the owners of the land they cultivated and possessed, but as in the United States, that ideal was seldom, if ever, upheld. After Mexico received its independence in 1821, it continued the practice of giving out land grants.

Several land grants were awarded in what is now southern Arizona. There was general peace between the Spanish and the Apache between 1790 and

1821, the date of Mexico's independence, and during that period there was Spanish movement northward. Expansion continued northward during the Mexican period, in spite of Apache raids, and many Arizona land grants were given during the Mexican period. Most of these land grants were designed for stock raising and, consequently, were very large.

The typical land grant in Arizona was 4 square leagues, or about 17,350 acres (7,027 ha.). All but one of the land grants in Arizona were long and narrow, to control both sides of a stream. By controlling the water, a grantee had mastery of much of the surrounding public domain. An owner was expected to use and occupy the land, and if it were ever abandoned for three years or longer, title was lost and the land again became public domain. Exceptions were made to this rule, however, as in the case of Apache raids. It was the government's responsibility to protect its citizens from the Indians, and if it could not do so, the settlers could not be expected to hold the land. If a grantee were forced to leave the land because of Indian threats, even for much more than three years, he still held legal title to the land. Mineral rights were never included in a land grant; all minerals belonged to the government.

After the American acquisition of Arizona, the Americans agreed to respect the land grants that were valid. The Mexican government was, of course, insistent that the rights of its citizens north of the new border be respected, and the Americans quickly agreed. The Americans established the Court of Private Land Claims, made up of five judges, to decide the land grant cases, and it continued in existence until 1904. People who were unhappy with a decision had the right to appeal to the U.S. Supreme Court. If there were a valid title, the grant was confirmed, if not, the grant was declared invalid and the claimant had no land. The most important criteria were that grants had to meet the requirements of Spanish or Mexican law at the time they were made and that the titles had to be legally maintained. Written evidence was required, and the grantee of the land had to provide that evidence. In Arizona a total of 11,323,108 acres (4,585,859 ha.) was claimed, but one of the claims, the Peralta Grant, was for a particularly large amount of land and was fraudulent. Only 837,680 acres (339,260 ha.) were seriously claimed and a total of 116,540 acres (47,199 ha.) were confirmed (Map 6.1). Most of the confirmed land very quickly left Mexican hands and was sold to Americans. Examples of land claims that were not confirmed are the Tumacacori Grant and the Peralta Grant.

The Tumacacori Grant was probably the oldest in Arizona, and it had been given to the mission. However, the mission was abandoned prior to 1820, though exactly when cannot be established, and in 1844 the grant was purchased at public auction by Manuel Maria Gándara, a powerful political figure in Sonora. He established a ranch there, but it was abandoned by 1859. In 1878 Gándara sold the claim, along with another, to two Californians, and they petitioned to have the claim confirmed. In 1891, however,

LAND GRANTS IN ARIZONA

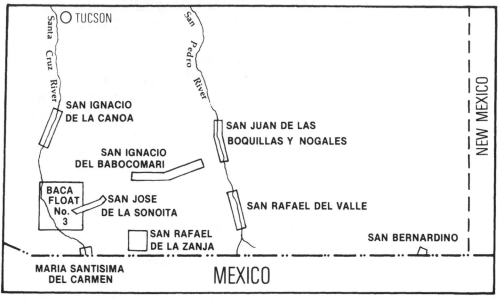

MAP 6.1

the Court of Private Land Claims found imperfections in the title. Since the Tumacacori Mission had been abandoned for many years, the land had reverted back to the public domain and the treasurer of Sonora had had no right to sell it. Therefore, Gándara, and all others who later claimed the land, had no legal title, and the grant was declared invalid.

The Peralta Grant was an outright fraud perpetrated by a convicted forger, James Reavis (Map 6.2). Reavis forged many papers and slipped them into the archives of Spain and Mexico. He then claimed 12,740,000 acres (5,159,700 ha.) of land in New Mexico and Arizona for his wife, who was apparently ignorant of his illegal actions. The case was taken before the Court of Private Land Claims in 1891, and Reavis almost got away with his scheme – indeed the Southern Pacific Railroad had already paid him to cross his land. But investigators in the foreign archives quickly recognized the forgeries, and Reavis was jailed.

The Baca Grant was a very unusual grant, and it resulted in the removal of 200,000 acres (81,000 ha.) of Arizona land from the public domain. The Baca Grant was established in 1835 when the Mexican government granted 500,000 acres (202,500 ha.) of land in New Mexico to Luis Maria Baca. In 1840 the city of Las Vegas, New Mexico, was also granted a large tract of land by the Mexican government, but it overlapped the Baca Grant. Both claims were accepted by the U.S. court, but the Baca heirs, in lieu of the land originally given, were allowed to choose blocks of land elsewhere, not to exceed five tracts. They chose five tracts of land, called Floats, of 100,000 acres each (40,500 ha.), two in New Mexico, one in Colorado, and two in Arizona (Map 6.2). The two Baca Floats in Arizona were No. 3 in Santa Cruz County and No. 5 in Yavapai County; both have been sold several times. Baca Float No. 5 is still largely intact, and it remains a working ranch in a very isolated section of the state.

American Land Claims

The first Anglo Americans to begin legally acquiring land in Arizona were miners. Near Prescott, miners organized the Walker and Pioneer Mining Districts in 1863, the first in the state. The secretaries of these organizations recorded the mining claims and enforced the rules. On May 10, 1872, Congress passed a law that recognized these mining districts as legal bodies, and

BACA FLOATS IN ARIZONA AND THE PERALTA CLAIM

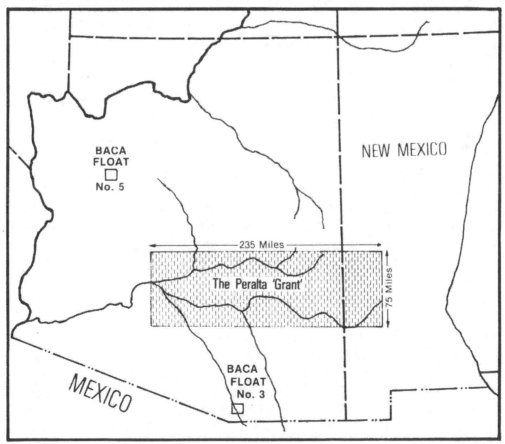

MAP 6.2

they could continue to make their own rules as long as they did not conflict with national laws. Thousands of mining claims were made in Arizona, but few were patented because the prospectors were often without funds to pay the required fees and also the prospectors quickly drifted off to other claims. Therefore, very little land was actually patented and removed from the public domain by the mining interests. The land that was acquired, however, was very important to the state because of the wealth it generated and because it encouraged immigration.

Several laws were passed by which citizens could acquire land cheaply. Even before the acquisition of Arizona, the Preemption Act of 1841 allowed people to settle an area on the frontier prior to any surveys. They would be "squatters" until they could acquire the land legally after surveying had extended to their land. The Homestead Act of 1862 allowed individuals to acquire 160 acres (64.8 ha.) for a small fee. The homesteader had to live on the land five years, and then he received, free of any further costs, full title to the land; or he could buy the land at a small cost ($1.25 or $2.50 per acre) after living on it for only six months. About 80 percent of the homesteaders in Arizona used the latter technique to acquire their land. The Timber Culture Act of 1878 allowed persons to acquire up to 640 acres (259 ha.) of desert land, but the owners had to have water and the means to irrigate the land. Under all of these acts only U.S. citizens could receive a patent to the land. Persons intending to be citizens could begin the process of homesteading, but they had to be citizens by the time the patent was awarded. In the Florence area along White's Ditch, for example, half of the claimants were Mexican Americans, but only half were citizens and thus able to acquire land under the Homestead Act.

There were many ways these acts could be abused, and they all were. Cattlemen would preempt water holes, or acquire land in long narrow strips along the streams, and thus control grazing on public land. Family members or employees would often attempt to use the Desert Land Act to acquire land for a "cattle baron." The Desert Land Act was particularly vulnerable to fraud because it had no requirement for local residency. Speculators paid individuals in distant cities for the use of their name to file on land, and those individuals then signed over their rights so that the speculators received the titles to the land. In these and other ways, large holdings were often accumulated and a great deal of land was monopolized.

Before any land could be acquired, it had to be surveyed. Surveying began in Arizona in 1867, though it wasn't until two years later that a land office was opened and people could begin to legally acquire land. Surveying began at the juncture of the Salt and Gila rivers, a point that had been located by the Boundary Commission after the Mexican-American War. All of Arizona, except for a small portion of the Navajo Reservation, was surveyed from this point. Surveying was done quickly in areas where agriculturists were settling, and within two years most of the Salt River Valley had been surveyed. Soon after that, surveying was done along the Gila. Each of the surveys that followed was in an area that was already somewhat settled by people anxious to acquire legal title to their land. This was particularly true in the case of farmers and ranchers who had settled near mining districts in order to have a market for their produce. These people wanted their status changed from squatters to bona fide homestead claimants, and the area around Prescott, for example, was surveyed as early as 1871. The surveys were not particularly accurate, and one can still see many discrepancies in the cadastral maps. The problem was that contracts were awarded to "qualified" surveyors, and they were paid for the amount of surveying accomplished, not on the quality of the work.

One final way in which large tracts of land were removed from the public domain was through railroad grants. The granting of government land to help defray construction costs of transportation lines was an old tradition; such grants were first made in the 1700s to aid turnpike builders, and in the 1830s grants were made to canal builders. Therefore, there was a legal precedent for granting land to the railroads, plus it had popular backing, and the first railroad to receive such a grant was the Illinois Central in 1850.

In 1866 the Atlantic and Pacific Railroad Company was authorized to build a rail line across Arizona Territory along the 35th parallel. The company was granted a right-of-way of 100 ft. (30.5 m.) along the track, additional land for such things as machine shops and depots, and alternate sections of land (640 acres [259 ha.] each) for 40 mi. (64.4 km.) on each side of the tracks. The company was required to finish the project by 1878, but it went bankrupt in 1873, and the rail line and its grant rights were acquired in 1876 by a new company, the St. Louis and San Francisco Railroad (known as "Frisco"). Four years later the Atchison, Topeka and Santa Fe bought a half interest in the enterprise, and under this new ownership the line across Arizona was completed in 1884. Although not completed in the allotted time, the Atlantic and Pacific Railroad was not forced to relinquish its land grants between the Rio Grande and the Colorado River, and the company received over 11.5 million acres (4.657 million ha.), with about 7.69 million of those acres (3.114 million ha.) in Arizona (Map 6.3). The Atlantic and Pacific was liquidated in 1894, the Santa Fe received this line (the Frisco took tracks elsewhere), and the two parent companies divided the remaining land.

The Texas and Pacific Railroad was awarded a similar grant along the 32nd parallel through southern Arizona, but the rail lines were never laid. The Southern Pacific extended its line through Arizona prior to 1882, and later bought a controlling interest in the Texas and Pacific Railroad, but Congress declared forfeiture of the lands granted, and the Southern Pacific in Arizona received only the right-of-way for its tracks, 9,902 acres (4,010 ha.).

PRIVATE LAND IN ARIZONA

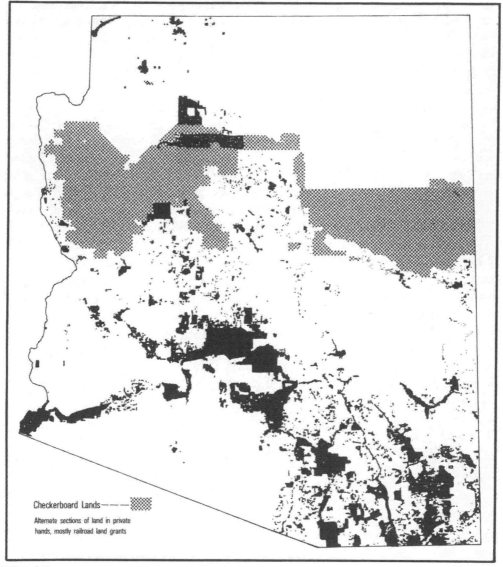

Checkerboard Lands————▨

Alternate sections of land in private
hands, mostly railroad land grants

MAP 6.3
Source: Office of Economic Planning and Development, Map, "Public Land Ownership in
Arizona, 1974."

Overview of Arizona Land

Not much of Arizona was actually removed from the public domain, and today only 17.5 percent of the state is in private or corporate hands (Map 6.3). The rest of the state consists of Indian reservations (26.7 percent), state land (12.3 percent), and other federally owned land (43.5 percent).

The federally owned land is scattered throughout the state (Map 6.4). Nonreservation federal land is governed by various agencies; for example, the national forests, national parks and monuments, and the military lands are all under different federal agencies; in addition there is the land controlled by the Bureau of Land Management.

The state owns about 8.8 million acres (3,561,248 ha.) of land, which is more than one and one-half times the size of Maryland. The state land is not in one block but is scattered across Arizona, and it was, of course, acquired when Arizona became a state. The state was entitled to four sections (1 sq. mi. [2.59 sq. km.] each) of each township (36 sq m. [93.2 sq. km.]), but some of the land already belonged to others – such as the Indians, the forest service, and so forth – so the state could then choose other land in its place. The state has yet to select about 170,000 acres (68,850 ha.) and will do so some time in the future.

The state sold some of its land, but there were few buyers. By law, grazing land had to be sold for at least $3 an acre and agricultural land at $35, and no one seemed to want to buy it, particularly the grazing land. If one bought grazing land, the taxes were more than the leasing fees, so ranchers just leased rangeland. The situation began changing in the 1960s when thousands of people began to move to Arizona. A great deal of land was sold, mostly to speculators and corporations, but the state began to retain its land so that, as it increased in value, the state could sell the land when it was more valuable. This practice is for the good of the state and all Arizonans and should be continued, or even better, the law should be changed so that the land would remain under state control for the benefit of all.

Having a large amount of land in state and federal hands benefits the average citizen, because the citizens have free access to huge portions of Arizona, and a few wealthy individuals and corporations cannot prevent them from hiking, hunting, or camping that land. For example, a rancher who leases state land for range purposes cannot keep out law-abiding citizens. Having a great amount of land in state and federal hands also means that the state has the opportunity to direct and control the growth within its boundaries.

However some very large blocks of land in the state are owned by individuals or corporations, and the two largest landowners own more than 11 percent of the state's private land. The single largest landowner is Tenneco West, a subsidiary of Tenneco, which owns the largest ranch in the state, the Diamond A Ranch north of Seligman, and a smaller ranch in the southeast, for a total of 604,000 acres (244,620 ha.). The company also leases other land

FEDERAL LAND IN ARIZONA

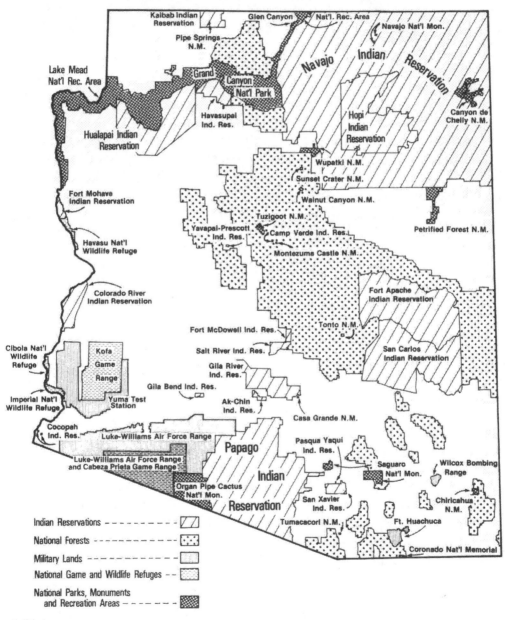

Kaibab Indian Reservation
Pipe Springs N.M.
Glen Canyon
Nat'l. Rec. Area
Navajo Nat'l Mon.
Lake Mead Nat'l Rec. Area
Navajo Indian Reservation
Grand Canyon Nat'l Park
Havasupai Ind. Res.
Hopi Indian Reservation
Canyon de Chelly N.M.
Hualapai Indian Reservation
Wupatki N.M.
Sunset Crater N.M.
Walnut Canyon N.M.
Fort Mohave Indian Reservation
Tuzigoot N.M.
Yavapal-Prescott Ind. Res.
Camp Verde Ind. Res.
Petrified Forest N.M.
Havasu Nat'l Wildlife Refuge
Montezuma Castle N.M.
Colorado River Indian Reservation
Fort Apache Indian Reservation
Cibola Nat'l Wildlife Refuge
Kofa Game Range
Fort McDowell Ind. Res.
Tonto N.M.
San Carlos Indian Reservation
Salt River Ind. Res.
Imperial Nat'l Wildlife Refuge
Yuma Test Station
Gila River Ind. Res.
Gila Bend Ind. Res.
Cocopah Ind. Res.
Ak-Chin Ind. Res.
Casa Grande N.M.
Luke-Williams Air Force Range
Papago
Pasqua Yaqui Ind. Res.
Wilcox Bombing Range
Luke-Williams Air Force Range and Cabeza Prieta Game Range
Saguaro Nat'l Mon.
Indian
Organ Pipe Cactus Nat'l Mon.
San Xavier Ind. Res.
Chiricahua N.M.
Reservation
Tumacacori N.M.
Ft. Huachuca
Coronado Nat'l Memorial

Indian Reservations -----------
National Forests -----------
Military Lands -----------
National Game and Wildlife Refuges --
National Parks, Monuments and Recreation Areas -----------

MAP 6.4

and controls a total of 929,454 acres (376,429 ha.). Tenneco West has no cattle, but it leases its land to a large cattle company headquartered in Phoenix. The New Mexico–Arizona Land Company is the second largest landowner, with 461,482 acres (186,900 ha.). It is a public corporation, and 51 percent of the stock is owned by the St. Louis and San Francisco Railroad (which acquired the original land grant rights of the old Atlantic and Pacific Railroad Company and then merged with the Santa Fe). The New Mexico–Arizona Land Company leases its land, which is between Holbrook and Springerville, to local ranchers. The third largest landowner is the Aztec Land and Cattle Company, which is owned by New England interests. (It is the old "Hash Knife Outift" of Zane Grey's novels.) This company's land, as well as that of the New Mexico–Arizona Land Company, is "checkerboard" land once owned by the Santa Fe (Map 6.3). The Santa Fe still owns about 124,000 acres (50,220 ha.) in Arizona and is the fifth largest landowner. The fourth largest land-owner is the Greene Cattle Company, which owns 165,000 acres (66,825 ha.), including the Baca Float No. 5 in Yavapai County. Much of the private land in Arizona is neither owned nor controlled by Arizonans. Also, most of the large ranchers do not own much of the land they use, but lease it instead.

Landholdings for agricultural purposes are much smaller but much more valuable on a per-acre basis. The largest farm is near Gila Bend, and it is owned by Northwest Mutual Life Insurance Company, which owns 72,000 acres (29,160 ha.) of land in Arizona. There are several other agricultural enterprises that farm between 5,000 and 10,000 acres (2,025 and 4,050 ha.), many between 5,000 and 1,000 acres (2,025 and 405.0 ha.), and a great many smaller farms. Of some concern to Arizonans is the fact that foreigners – such as Canadians, Japanese, and Arabians – are investing in Arizona farmland, and by early 1980 foreigners owned over 70,000 acres (28,350 ha.) of Arizona cropland. The total amount of agricultural land in the state in 1977 was 1,327,670 acres (537,706 ha.). Most agricultural land is privately owned, but some of it is on Indian reservations.

Mining and other industries do not use much land. According to the Arizona Mining Association, less than 200,000 acres (81,000 ha.) in Arizona are occupied by mines, mills, smelters, waste disposal areas, and accompanying roads – about one-fourth of 1 percent of the state. Manufacturing uses very little land, since a million gadgets can be made in a very small area. Mining and manufacturing, however, are the two leading industries in the state, producing a great amount of wealth but using only a small amount of land. What is important to the state, therefore, is not how much land is used by whom, but how it is used and who controls it.

WATER

Water is a critical and important factor for anyone living in a desert environment such as Arizona. Arizona receives very little precipitation, and what it does receive is not evenly distributed across the state (Map 2.4). The densely populated portions of the state are in the drier areas and, thus, dependent on moisture falling in other regions being brought to them. It is also common in desert areas that there is a great amount of variation in rainfall from one year to the next, and this is true in Arizona; an area that "normally" receives 10 in. (25.4 cm.) of rain a year may receive 15 (38.1 cm.) one year and 5 (12.7 cm.) the next. A variation of 10 in. (25.4 cm.) may not seem like much to someone from a humid area, but in Arizona such a variation can mean success or failure, particularly in farming and ranching. However, the total average rainfall in Arizona has probably changed very little, if any, since the arrival of man over 20,000 years ago.

The first humans in Arizona and all nonagricultural Indians needed little more than enough water for drinking, so the lack of water did not affect them greatly since their needs were slight. Long before the arrival of the whites, the Indians in the Southwest had tales of evil spirits inhabiting the water holes, so the Indians always lived some distance from water. These tales were probably based on the fact that living close to a spring could cause pollution and contamination of the water, which could result in sickness.

The agricultural Indians needed large, dependable supplies of water, but all of these Indians lived in the drier sections of the state, and, therefore, they were dependent on some sort of irrigation. The Hohokam in particular lived in some of the hottest and driest portions where the water needs were greatest. The Hohokam built small brush and rock dams and long and extensive canals in order to divert water to the fields, but they faced many problems. Flooding must have been a major problem, and in spite of modern technology it is still a problem in the Salt River Valley. Another problem was waterlogging; repeated irrigation would cause the water table to rise almost to the surface, which would kill the plants. Alkali and salt concentrations on the surface also killed some crop plants; in fact that last problem forced the abandonment of much of the land west of Tempe after it had been farmed for only 25 years during the early American period. Whites solved these problems by pumping (the first pumps in the Salt River Valley were steam driven and were used to lower the water, not to irrigate the crops), digging deep drainage ditches, placing tile beneath the surface to carry off water to drainage ditches, and flushing the soil whenever possible. The Hohokam did not have the understanding or the techniques to solve the problems, and they probably survived by abandoning used land and extending the canals to new areas.

The Spanish in southern Arizona also had need for irrigation, and they built a few small irrigation canals along the Santa Cruz River. They certainly knew the techniques of canal construction, but they only copied and expanded a few Indian canals at the missions and at agricultural sites. The land was not heavily used, so there was no need for the Spaniards to build large or elaborate irrigation systems.

The first Anglos in southern Arizona knew very little about irrigation, but they soon learned. Some of the first irrigation by Anglos was in the Salt River Valley. The men simply pooled their labor to build small dams and dig canals that generally followed the old Indian canals, especially on the south bank (Figure 6.1). The men who did the work then shared the water. There were soon too many canals competing for water, and trouble developed; the people involved did not have a tight-knit society, nor did they have the laws or means to enforce justice. There was soon armed conflict and the destruction of upriver dams during times of water scarcity (Figure 6.2). Even in good years there was hardly enough water for all, and in years of scarcity, such as from 1897 to 1899, the river tended to dry up, agriculture failed, and violence flared over water ownership.

Arizona Water Laws

Americans in the East used a system of water laws, which had evolved from the old Anglo-Saxon laws, called riparian doctrine. The riparian doctrine gave a landowner many rights if a stream touched or flowed through his property. He could do with the water as he saw fit – such as use it, divert it, sell it, prevent others from using it, and so forth – but there were restrictions, particularly in that water could not be taken from a watershed, or stored, and it had to be shared equally by everyone whose land touched the stream. Americans took these laws west with them, but these laws and ideas just did not work in a desert environment. The Americans found that the earlier Mexican settlers used a completely different set of laws, which had evolved from the dry areas of southern Europe and northern Africa. Under this system a landowner could live a long way from a stream, yet divert all of the water to his use, and leave the stream dry below his canal. The Americans arriving in the West did not understand these laws and did not consider them fair; but with time, they realized these laws worked better in a desert than did the riparian doctrine. It was better, for example, for one person to have enough water for crops than it was for everyone to share the water and no one have enough (Figure 6.3).

The water law that developed in Arizona pertaining to surface water is known as the Law of Prior Appropriation and Beneficial Use. It evolved from the older Spanish water laws used in New Mexico and has two distinct parts: (1) the doctrine of appropriation and (2) the doctrine of beneficial use. The latter part of the law implies that water should be used for the good of all, and the courts are always having to interpret just what is "beneficial use." The doctrine of appropriation implies that the first settler to use the water has the first right in times of shortage. This use could be for almost anything – for agriculture, turning a mill, or mining – as long as the use is beneficial. Newcomers might try to use the water, but the courts have upheld the rights of the first settlers, and newcomers can only get water if

FIGURE 6.1. Early canal construction in the Salt River Valley. Courtesy, the Salt River Project.

FIGURE 6.2. An armed horseman guarding a canal headgate in the early Anglo period of the Salt River Valley. Courtesy, the Salt River Project.

FIGURE 6.3

DIFFERENCES IN RIPARIAN AND PRIOR-APPROPRIATION WATER LAWS

Riparian Doctrine:	Appropriation Doctrine:
1) Acquired by ownership of riparian land.	1) Acquired by compliance with statutory procedures, and not dependent on land-ownership.
2) Does not depend on water use for creation or continued existence.	2) Depends on water use for creation and continued existence.
3) Entitled to co-equal share of water supply.	3) Strict priority between senior and junior appropriators.
4) Entitled to uncertain quantity of water.	4) Limited by terms of permit to maximum amount of water use, purpose, and time and place of use.
5) Can use water only on riparian land and only within the watershed.	5) Can use on non-riparian land and outside the watershed if approved by permit.
6) Generally cannot store water for future use.	6) Permit can allow reservoir construction and water storage for future use.
7) Can settle disputes between claimants only by court action.	7) Can often settle disputes between claimants by administrative action.

Source:
Otis W. Templer, "Texas Surface Water Law," Historical Geography 8 (1978): 12.

there is a surplus. The water has to be used continually, or regularly, or the rights could be lost. If the water is not used by the one who has the right to it for approximately three years, but is used by newcomers, those newcomers have the better right to the water based on the year they began using it. With these water laws farmers can build dams and divert water to their fields and know where they stand in regard to priority of water use. If their rights are very old, for example, they are guaranteed water, but if their rights date to more recent years, they might receive water only in times of plenty. Under this system, the water rights belong to the land, not to the individuals, and water belonging to one piece of land cannot be sold for use elsewhere. These surface water (or gravity water) laws are still very much in force in Arizona.

But large changes in water use are coming about as a result of Indian demands for water rights. The farmers in Arizona have long argued that the Indians have as much right to the water as everyone else. If the Indians were irrigating their fields before the arrival of the whites, then their water rights for that amount of water are much better than those of the newcomers. If the Indians were not using the water before the whites arrived, such as the Apache hunters and gatherers, then their rights to the water date from the time they started using it. However, these concepts have not been upheld. According to the federal government, reservations reserve water as well as land for the Indians, enough to irrigate as much land as is practical, and the water laws of the state do not apply to the Indians. This concept, the Winters doctrine, was clearly stated by the U.S. Supreme Court in 1908 when it said that streams running through or adjacent to the reservations contained water reserved for the Indians and that this water was exempt from appropriation by non-Indians. This doctrine means that if an Indian reservation was established in 1870 but only began using water in 1970, in time of water shortage those Indians' water rights are better than the rights of any land on which irrigation began after 1870. The 1910 Kent decree (named for Judge Edward Kent) decided the ranking order in which Salt River Valley landowners receive water, but that decree does not deal with upstream users and cannot be extended to that area. The Apache in eastern Arizona have begun to build a large number of stock tanks, dams, and irrigation works on their reservations, and the traditional water users in the lower valleys have no recourse. This Indian use will someday bring changes to water use in central Arizona.

The laws relating to groundwater use are very different, and they derive from English or common law. If there is an underground stream, and it can be proven to exist, it can be appropriated, but this rarely occurs. Otherwise, groundwater is not subject to appropriation, and the first person to sink a well has no more rights than any other well owner. There has been such an abuse of groundwater in Arizona that in the 1950s the courts began to give strong consideration to the principle of "reasonable use." But the problem is one of ownership: who owns the groundwater that is pumped on one person's property when obviously much of it comes from beneath his neighbor's land?

The first large wells and pumps in the Salt River Valley were primarily to lower the water table, not to irrigate the land, and most of the serious pumping of water for agricultural purposes did not begin until about World War II. Prior to World War II there were about 500,000 acres (202,500 ha.) in Arizona under cultivation, but by the late 1960s that figure had risen to 1,300,000 acres (526,500 ha.), and almost all the increase was as a result of pumping. By 1957 there were 3,500 irrigation wells in the Salt River Valley and the lower Santa Cruz Basin. Many of them were very large wells, pumping up to 5,000 gal. (18,925 l.) per minute, and some of them were quite deep, up to 2,000 ft. (609.6 m.). There seemed to be a race by the farmers to see who could pump out the most water. Those people such as the Indians who did not pump, either for economic or ecological reasons, came to realize that much of the water beneath their land had been removed by their neighbors. The farmers were literally mining the water, something that had been deposited during the Pleistocene epoch, which ended about 10,000 years ago.

The Ground Water Act of 1948 attempts to protect the groundwater. It prohibits any future drilling of irrigation wells in those areas designated as "critical groundwater areas," but existing wells can be deepened as the water table drops. Critical groundwater areas are those areas deemed not to have a safe, reasonable supply (Map 6.5). All areas so designated are areas of immense groundwater reserves, because it is in such areas that wells have been drilled and there has been intensive pumping. According to the law, wells can be drilled for any purpose other than for irrigation, such as for livestock, home use, and industry. If such a well produces over 100 gal.

DESIGNATED CRITICAL GROUNDWATER BASINS

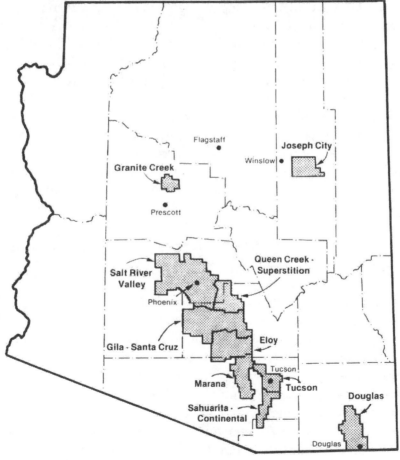

MAP 6.5
Source: Arizona Water Commission, "Arizona State Water Plan, Phase I: Inventory of Resource and Uses" (July 1975), p. 66.

(378.5 l.) per minute, a permit has to be acquired, but permits are automatically given, and the system is only used to keep track of all the new wells. In areas outside a critical groundwater area, any kind of a well can be dug, and the water can be used for any purpose.

Much of the groundwater used in Arizona today is not being used wisely. Many of the farms are run by corporations owned by large companies or wealthy individuals, and many of those farms lose money or barely break even. The farms are then used as a tax write off, but they are also wasting water that should belong to future generations. The major group to benefit are the Japanese who buy the major product produced – cotton – manufacture clothes from it, and then sell the clothes back to Americans at greatly increased prices. One may argue that the farms are needed to produce food, but only about 5 percent of the farmland in Arizona is in vegetables and only about 7 percent in citrus; the rest is mostly in cotton and alfalfa.

The groundwater law is now being changed, mainly as a result of pressure by President Carter, who agreed to provide funds for the Central Arizona Project only if the state established new groundwater laws. The old laws applied mostly to agricultural interests, but since the mid-1950s towns have grown, and today urban needs must be recognized. A state bill (No. 1391) was passed in 1977 that initiated some of the needed changes. It details a system for assessing damages if someone's water table is lowered by a neighbor's pumping. It allows cities to transfer water from one area to another, and it permits them to buy agricultural land, retire it, and transport the pumped water to the cities. These changes are needed, and it is to be hoped that when the groundwater laws are completely revamped, it will be done in a way that will benefit the majority of the Arizonans as well as future generations.

Areas of Groundwater Use

There is not much groundwater on the Colorado Plateau of northern Arizona. The rocks in that area lie flat, and some of the layers, especially the sandstones, are aquifers, or water bearing (Figure 6.4). But there are many problems with these aquifers; they do not hold much water, water does not travel through them easily, recharge is very slow, and some of the water is not very good because it is highly mineralized. There is enough groundwater in the area for domestic and livestock use, but the wells are usually small (pumping up to 200 gal. [757 l.] per minute), and they are often very deep, which makes them expensive to operate. Exceptions to this generalization are those areas where there has been faulting and fracturing of the rock. In those areas the fracturing allows water to accumulate, and there is greater permeability, both vertically and horizontally. In such areas wells can produce good amounts of water, like several near Flagstaff or the one at St. Johns that produces up to 1,800 gal. (6,813 l.) per minute.

ROCK FORMATIONS AND WATER-BEARING STRATA IN NORTHERN ARIZONA

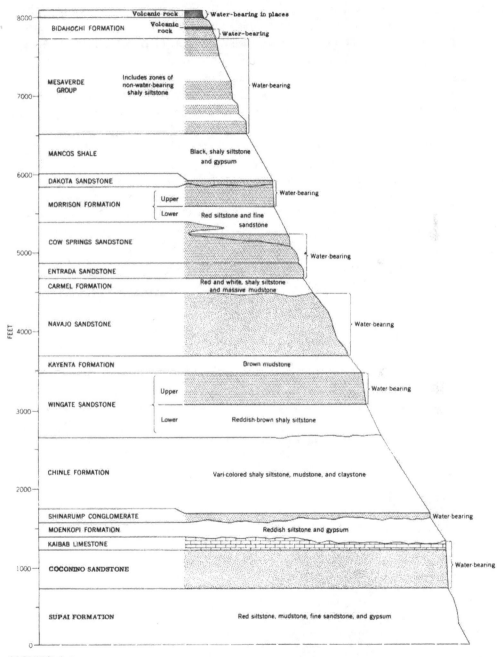

FIGURE 6.4
By permission from *Arizona—Its People and Resources*, Faculty of the University of Arizona, (Tucson: University of Arizona Press, 1972).

On the other hand, the geological formations of southern Arizona are very favorable for water accumulation. The alluvial basins in the Basin and Range Province are very deep – they could easily be 3,000 to 5,000 ft. (914.4 to 1,524 m.) deep – and they contain a great deal of water (Figure 6.5). It is not known how much water is in these basins or how long it will last. The quality of the deep water is also unknown, though almost all of the basins now produce good-quality water that is free of salt. It is in these large basins, with their vast amounts of water, where there has been extensive pumping and, consequently, large drops in the water table.

Southern Arizona is particularly dependent on groundwater. Phoenix and the Salt River Valley have both surface water and groundwater, but many of the cities, towns, farms, and mines in southern Arizona are completely dependent on groundwater that cannot last forever. Tucson is the largest city in the world to supply all of its water needs from groundwater, and thus, that city is greatly concerned about the overall water policy and the changes in the law.

The water situation along the California border is different. There can be no pumping of well water in the Salton Sea area of California because the underground water there is mostly salty. The situation is essentially the same along the California border, though many good wells do exist. A high water table can often become a nuisance in that area, because so much water is needed by the crops that some of it soaks in to raise the water table.

IDEALIZED PROFILE OF EARTH CRACK DEVELOPMENT

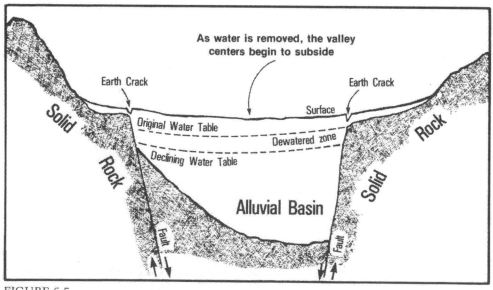

FIGURE 6.5

Problems in Groundwater Use

The use of groundwater causes some problems in southern Arizona where there is a great amount of groundwater but very intensive use. Too much water is being removed, and in many places the water table is dropping very rapidly. The water table is first lowered immediately next to the well, which results in a "cone of depression," but it does not take long for the entire water table to begin dropping. In some areas, the water table has dropped as much as 300 ft. (91.4 m.), and in Deer Valley north of Phoenix the water table dropped that much in only one year of intensive pumping. The greatest drop in the water table is in two areas: (1) the Salt River Valley and (2) the lower Santa Cruz Basin around Maricopa and Casa Grande (Map 6.6).

CHANGE IN GROUNDWATER LEVELS, 1940-76

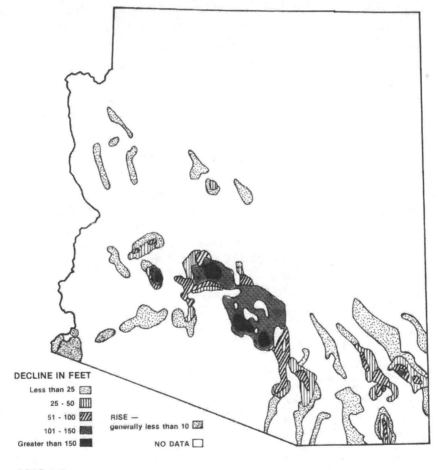

DECLINE IN FEET

Less than 25

25 - 50

51 - 100   RISE —

101 - 150   generally less than 10

Greater than 150   NO DATA

MAP 6.6
Source: U.S. Geological Survey, "Water Resources Investigations in Arizona," (1977).

In both areas the water table has dropped as much as 300 ft. (91.4 m.). The only place in Arizona where there is intensive pumping but only a relatively low drop in the water table is the Safford area of the upper Gila River valley, because there is almost as much recharge of the water table from the Gila River, which is always carrying water, as there is pumping. In most areas of the state there is a great difference between the amount of water pumped and the amount of recharge; although 52.1 percent of water used in the state is groundwater, only 6.3 percent can be considered naturally recharged. The overdraft is thus 45.8 percent of all water used, or 2.2 million acre-feet per year (Figure 6.6), and 88 percent of all pumped water is overdraft. (An acre-foot of water is the amount of water needed to cover 1 acre [0.405 ha.] of land 1 ft. [30.48 cm.] deep. One acre-foot equals 1,233.5 cu. m.)

The most obvious problem of all this pumping is that, if there are no changes, the wells shall be pumped dry. In many places there is enough water to last 100 years, but that is not really a very long time, and changes must be made long before then. Bringing Colorado River water into central Arizona will help, but it will only slow down the overdraft. Arizona's greatest hope of permanently solving the problem is in the new water laws, and the one industry that will probably suffer is agriculture, an industry that

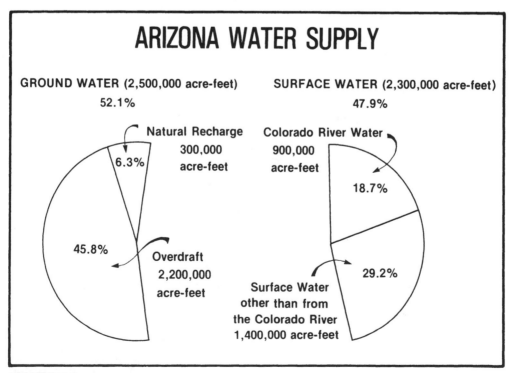

FIGURE 6.6
Source: Arizona Water Commission, *Arizona State Water Plan, Alternative Futures, Phase II* (1977).

now uses almost 90 percent of the groundwater. Indeed, the dropping of the water table is almost always associated with agriculture.

Another problem is that as the groundwater is withdrawn, the earth tends to settle. The earth has dropped more than 8 ft. (2.4 m.) in the Santa Cruz Basin and over 4 ft. (1.2 m.) in the Salt River Valley. This is not an unusual situation, and it has occurred in many parts of the world where there has been an intensive pumping of water or oil. Two things happen as the earth settles. First, as the surface drops, dirt fills in some of the tiny spaces once occupied by water, so the potential for water once again entering the water table is lessened. Second, tremendous stress is placed on the edges of a valley as its center drops, and large earth cracks develop parallel to the mountains (Figure 6.5). These cracks can be 30 ft. (9.1 m.) wide and very deep, and because of them valuable land and any buildings built over them are destroyed (Figure 6.7). Perhaps these cracks are nature's way of repairing the harm done by intensive pumping, because after a rain, washes coming out of the mountains carry water into the earth cracks and the water disappears underground. One theory argues that pumping has caused crystal uplift; for example, the removal of 95.7 trillion lb. (43.4 trillion kg.) of water from the lower Santa Cruz Basin between 1948 and 1967 caused the crust of the earth to rise 2.4 in. (6.1 cm.). It is certainly possible that the theory is correct.

Attempts to Conserve and Increase Water Supplies

Attempts are now being made to conserve water in Arizona. It must be remembered that only 5 percent of the water that falls in the Salt River Project watershed can actually be used in the valley's canals; the other 95 percent is lost, mainly to evaporation and transpiration. If only another small percentage can be saved, it will greatly increase the amount of water available for use.

One method of increasing the runoff is vegetation modification. This technique can be very useful, but it can result in trouble if not done carefully, and environmentalists usually object to it. Some proponents carry vegetation modification too far and want to remove all trees – such as willow, cottonwood, sycamore, and salt cedar – along a stream's course, because those trees use a great deal of water. But the trees do provide a valuable habitat for wildlife, and much of that habitat should be saved. The greatest benefits would come if chaparral were replaced by grass, because the water runoff on hillsides would greatly increase. The water would run off over a long period of time, and there would be fewer silting problems because the water would be cleaner. Chaparral removal would also benefit the cattle industry. Chemicals were once used in efforts to control the chaparral, but chemicals can be dangerous. In the late 1960s the U.S. Forest Service and/or the Salt River Project sprayed the herbicide Kuron near Globe, and a major suit now contends that the chemical caused physical af-

242

FIGURE 6.7. An earth crack east of Phoenix.

flictions and contributed to the death of at least one person. Natural methods, such as the use of fire and the grazing of goats, may prove to be the best methods of chapparal control.

Another method of conserving water is through groundwater recharge. Artificial recharge–pumping water underground–has been practiced for about 100 years in other parts of the world, and it holds promise for Arizona, though there are some economic, engineering, and legal problems–what is to prohibit one person from pumping out water just pumped in by a neighbor? Underground storage is an excellent way to store water, because there is no loss to evaporation and 100 percent of the water can be pumped out when needed. Many people think that all the water that flows over a dam and floods the washes is "lost" to Arizona, but quite a bit of that water soaks underground to recharge the water table. The Salt River flooded several times in the 1970s, and the water table near the river rose dramatically each time. After a flood in the winter of 1978–1979, the water table rose over 78 ft. (23.8 m.) on the east edge of Phoenix, and a rise of at least 90 ft. (27.4 m.) has been recorded on the west side where the water slows down and spreads out. All the water that is thus stored underground can be reclaimed later, and it will be much cleaner than when it entered the valley. It would be to the overall advantage of Arizonans if water were frequently released from the Salt River Project dams and a series of dikes were built in the riverbed in order to slow the water's flow and encourage percolation into the ground.

Converting salt water to fresh water is another possible way to increase Arizona's water supply, but conversion is a long-term solution, and there is no hope for its immediate use. It is very expensive to convert seawater to fresh, and the place where water is needed most, Tucson, is 2,500 ft. (762 m.) above sea level; pumping seawater that high and far would be very expensive. Brackish water does exist underground in southeastern Arizona, and conversion of that saline water to fresh water would be more practical.

There are many other methods of increasing Arizona's dependable water supply, and the Central Arizona Project, which will bring Colorado River water to central Arizona, will aid the water situation greatly. In the long run, however, it will be through conservation techniques and increasing the runoff that Arizona will be able to once again balance long-term water availability and water use.

Major Irrigation Projects

For agriculture to be a large and lasting enterprise in Arizona, major irrigation projects were necessary, and many have been carried out. The first irrigation project, and the one with the greatest impact on Arizona, was designed to control the Salt River, a necessity if agriculture were to succeed in the Salt River Valley. There always seemed to be too much or too little water; in some years the river almost dried up, and in others flash floods would destroy the early, small dams. Also, much of the water came when it could not be used. Many of the floods came in late summer after crops had been harvested, and the water of many early winter floods also went unused. In order to store water from one year or season to the next, a large dam was needed. This was too expensive an undertaking for local groups, so they turned to the federal government for help.

The Reclamation Act of June 17, 1902, was passed in order to allow the federal government to aid in carrying out irrigation projects, and the first such project started was along the Salt River. All of the little canal companies in the Salt River Valley were consolidated into the Salt River Valley Water Users' Association, so that the government would have to deal with only one agency. The Salt River Valley Water Users' Association bonded all the land within its bounds as security for a federal loan to build the dams and canals, and the work was started.

The first dam, Roosevelt Dam, was begun in 1905 and completed in 1912 (Figure 6.8). This dam can, and often does, hold 1,420,000 acre-feet of water. In order to divert water into canals on the north and south banks of the Salt River, the Granite Reef Diversion Dam was built in 1908 (Map 6.7). It is a

FIGURE 6.8. Roosevelt Dam. This picture illustrates how the water of Apache Lake, behind Horse Mesa Dam, backs up almost to the bottom of Roosevelt Dam. Courtesy, the Salt River Project.

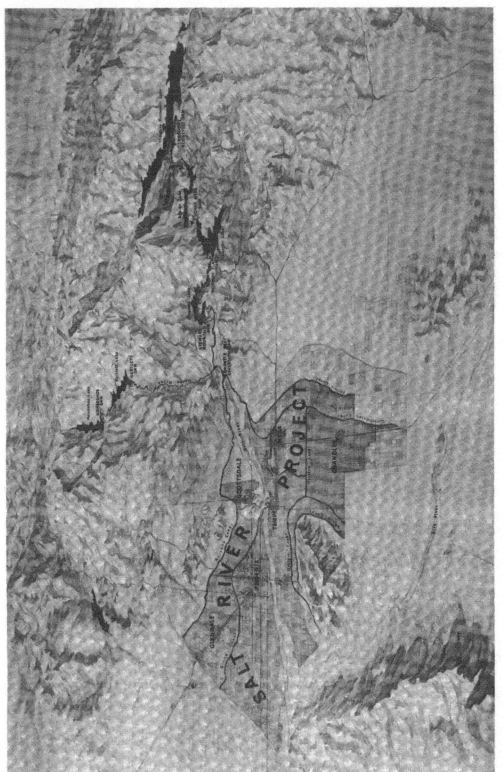

MAP 6.7. The Salt River Project and central Arizona. Courtesy, the Salt River Project.

low dam, only 29 ft. (8.8 m.) high, and it simply raises the water a few feet to divert it into the canals. In 1917, the government turned over the care, operation, and maintenance of the dams and canals to the Salt River Valley Water Users' Association, and the association was to repay the cost of building them to the federal government. Within eight years of completion, Roosevelt Dam had overflowed four times, so three other large dams were built on the Salt River below Roosevelt Dam: Mormon Flat Dam (1923–1925, capacity of 57,582 acre-feet), Horse Mesa Dam (1924–1927, capacity of 245,138 acre-feet), and Stewart Mountain Dam (1928–1930, capacity of 69,765 acre-feet).

Later two dams were built on the Verde River in order to control and conserve its water. Bartlett Dam (1936–1939, capacity of 178,185 acre-feet) was built with federal funds, but the Salt River Project agreed to pay 80 percent of the cost. Horseshoe Dam (1944–1946, capacity of 139,238 acre-feet) was financed by the Phelps Dodge Corporation, and in exchange that corporation received water credits for the water impounded behind the dam. In 1949 the city of Phoenix paid to have spillway gates added to Horseshoe Dam, and as a result Phoenix received increased water credits for its domestic water supply.

With the completion of the dams, the Salt River Project began supplying the Salt River Valley with about 800,000 acre-feet of gravity water per year (Map 6.8), and since their completion, the water level has gotten dangerously low only once; on September 18, 1940, there was only a three-day supply of water left when the rains came. There is also extensive pumping of groundwater, and about 400,000 acre-feet of pumped water is put into the canals each year (Figure 6.9), which means that about one-third of all the water in the Phoenix area canals is groundwater.

All of the dams on the Salt River were designed to produce hydroelectric power. Also, three hydroelectric plants were placed in the canals to utilize the power of the water as it dropped into the valley, and one of these plants is still producing electricity. Even before Roosevelt Dam was completed, power was being produced in order to help build the dam, and by 1909 power produced at Roosevelt Dam was being used in the valley, much of it for pumping water. Cities had been served by private utility companies, but because of the economics, the farmers had had no electric service. The Salt River Valley Water Users' Association began providing power to its members in the valley and truly entered the utilities business. The Salt River Project Agricultural Improvement and Power District was formed in 1937, with boundaries almost identical to those of the association. This organization is a particularly significant and powerful one, because it has all the rights, privileges, exemptions, and immunities granted to cities. In particular, it has made possible refinancing through tax-exempt municipal bonds. The Salt River Valley Water Users' Association and the Salt River Project Agricultural Improvement and Power District are, thus, two legal

SALT RIVER PROJECT WATERSHED AND IRRIGATED AREA

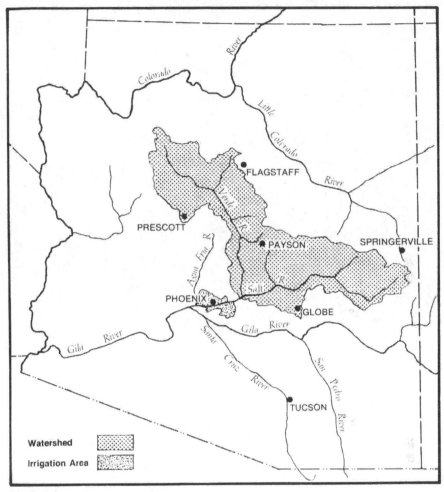

MAP 6.8

bodies, but they operate as one—the Salt River Project. By the early 1950s, the Salt River Project had begun building generating stations to produce power, since the power produced by the dams was insufficient, and it is now involved in almost all of the major power-producing projects in the Southwest. It also began buying power from other producers. The Salt River Project also built "pump-back" facilities at Mormon Flat and Horse Mesa dams; these facilities pump water back into the higher lakes at times of low demand, and the water is then released to generate power at times of peak demand.

The sale of power dominates the revenues of the Salt River Project and is

FIGURE 6.9. An electric pump, belonging to the Salt River Project, pumps groundwater into a canal.

now quite controversial. In 1977 it received over $305 million operating revenue from the sale of electric power, but only $5.4 million from water and irrigation. One not only clearly dominates the other, but also supports it. In 1978, approximately $16 million worth of power revenue was used to help reduce the cost of water for the farmers. Consequently, water is *very* cheap in the bounds of the Salt River Project. In 1978, the cost for water was $5 each for the first two acre-feet, $15 for a third acre-foot (obtainable when water stored in wet years is available), and $16 an acre-foot for a fourth and fifth acre-foot of groundwater. A farmer can thus receive a maximum of 5 acre-feet of water for an average cost of $11.40 per acre-foot. This is cheap by any measure; water in Tucson costs ten times that amount. The farmers are pleased with the situation, but the urbanities, who buy most of the electricity and thus subsidize the water, are not. However, the farmers are very powerful because of the way the Salt River Project was established, and they are not going to change the situation.

When the Salt River Valley Water Users' Association was established, the landowners pledged their land as collateral, and each acre (.405 ha.) pledged represents one share in the association, one vote in elections, and one part of the total debt. The system is known as "debt proportionate voting"; a farmer with 160 acres (64.8 ha.) has 160 votes, and the average suburbanite, living on one-fifth of an acre (.0810 ha.), has one-fifth of a vote, so the farmers easily elect other farmers to the Salt River Project Board, which establishes the policies. Thus, electric rates are kept high in order to keep the water costs down. Traditionally, there were 10 board members, but 4 "at large" members have been added. These 4 members are to represent urban interests, but they can still be outvoted by farming interests 10 to 4. A recent court ruling struck down debt proportionate voting in favor of a "one-person, one-vote" system, but that decision will be appealed to the Supreme Court. If it is upheld, the leadership and control of the Salt River Project will be in the hands of the urban residents in the Salt River Valley, and changes will quickly come.

There is also extensive irrigation agriculture along the Gila River. Along the upper Gila, there is a long ribbon of agriculture centered on the town of Safford. In that area 16 small irrigation companies provide the water, both gravity water and groundwater (which is high in soluble salts). But the major area of irrigation along the Gila River is in central Arizona, stretching from Florence onto the Gila River Indian Reservation, and this area of irrigation is known as the San Carlos Project.

Coolidge Dam is the only large dam to have been built for the San Carlos Project (Map 6.9), though there is a small diversion dam, the Ashurst-Hayden Dam, a few miles east of Florence. The Coolidge Dam, completed in 1928, has a capacity of 1.3 million acre-feet, but it was never more than 70 percent full until it filled for the first time in early 1979, over 50 years after it was built. The dam could actually have been made half its present size and still perform its intended function. The lake behind the dam, San Carlos Lake, is usually very low and very small; Will Rogers, who was present for the dedication in 1930, described it best when he said, "If this were my lake, I'd mow it." The dam was initially proposed and built to serve the Gila River Indian Reservation, because the whites at Florence, upstream from the reservation, were taking most of the river's water. The whites complained, arguing that the Indians did not need all that water, so it was finally agreed that the water stored behind the dam would be divided by whites and Indians. It was thought that the dam would provide water for 100,000 acres (40,500 ha.), but that much water just was not there. Eventually about 50,000 acres (20,250 ha.) were cultivated, mostly on land owned by non-Indians, but the entire area is largely dependent on groundwater. Today, a very extensive area south and east of the San Carlos Irrigation and Drainage District has been cultivated by using groundwater. This is known as the lower Santa Cruz Basin (the Central Arizona Irrigation District and the Maricopa-Stanfield Irrigation District), and at least 230,000 acres (93,150 ha.) in this area are used for agricultural purposes.

There are many smaller irrigation districts scattered across the state. Many of these other areas are dependent on groundwater, but a few have ac-

IMPORTANT DAMS IN ARIZONA

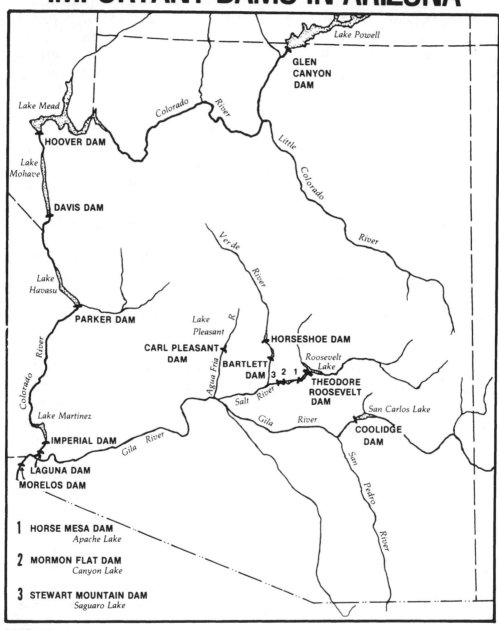

MAP 6.9

cess to some gravity flow water, such as along the Big Sandy River in western Arizona, near Wilcox and the Kansas Settlement in southeastern Arizona, near St. David along the San Pedro River, along the upper Verde and the lower Gila rivers, and many others. There are also many small irrigation districts around the bounds of the Salt River Project, and most of those are dependent on groundwater. They include the San Tan Irrigation District, Roosevelt Irrigation District, McMicken Irrigation District, Maricopa County Municipal Water Conservation District, Chandler Heights Irrigation District, Roosevelt Water Conservation District, and several others. There are also many independent farmers who own and operate their own wells (Figure 6.10).

Another major source of water for Arizona is the Colorado River, but there are problems concerning the use of this water. Much of the land on the Arizona side of the river is relatively high in relation to that on the California side. This makes it much easier for Californians to use the water, and they began using it at an early date. They have, thus, acquired "prior rights," but as the Colorado River is an interstate and a navigable stream, it comes under federal control, and water is distributed according to other laws. Division of such water is generally solved through the courts, usually a long and costly

FIGURE 6.10. A privately owned diesel pump. The 55-gallon barrel to the left of the pump drips oil on the shaft of the pump to keep it lubricated.

process for all involved.

Much of the California use is in the Imperial Valley (south of the Salton Sea) and the Coachella Valley (north of the Salton Sea). Much of that area is below sea level, and there used to be the danger that the Colorado would flow that way rather than to the Gulf of California. (That did happen in 1890–1891, when the river broke through its right bank and formed the Salton Sea.) Farming in those California valleys by using water from the Colorado River began in 1901; and the farmers, in an effort to get more water, made a small breach in the right bank of the river in the summer of 1905. That was a mistake, and the whole area would have been a vast lake had it not been for the Southern Pacific Railroad. The railroad realized it could make no money crossing or going around a lake, so it blocked the break and saved the area for agriculture. This area can use the water very easily, the water generates electricity as it drops into this low area, and irrigation water seeping underground maintains the level of the Salton Sea.

The way the water of the Colorado River is divided is a result of politics and is very complicated. Congress in 1921 authorized the formation of the Colorado River Compact Commission to negotiate water rights among the various states. The result was the Colorado River Compact, which was ratified by all concerned states except Arizona, by 1923. The agreement divided the region into two areas, the upper basin states (Utah, Colorado, New Mexico, and Wyoming) and the lower basin states (California, Nevada, and Arizona). The actual dividing point is not on the Arizona-Utah border, but at Lee's Ferry, a few miles into Arizona. The water from the Colorado is to be equally shared, and any future treaty giving Mexico a share of the water is to be borne equally by the two areas. It was estimated that the river annually carried 16.4 million acre-feet at Lee's Ferry, and the upper and lower basin areas were to receive 7.5 million acre-feet each. This estimate was a major error, for the Colorado seldom carries that much water. The estimates of the flow were made in wet years during the early part of this century (1899–1920), and such large flows are rare—the last probably occurred in the early 1600s.

The Boulder Canyon Project Act cemented much of the legislation and the water rights on the Colorado. The act authorized the construction of Hoover Dam and established a formula for water use on the lower Colorado: California was to receive 4.4 million acre-feet, Arizona was to get 2.8 million acre-feet plus all the waters of the state's tributaries, and Nevada was to get 300,000 acre-feet. Arizona refused to accept this division, realizing that it has a much longer border on the Colorado than does California. California, meanwhile, because of its topography and political power, continued to take the water it needed. The division of water established by the Boulder Canyon Project Act was legally finalized in 1964 by a U.S. Supreme Court decree. Arizona had to accept this division of water.

Arizona finally ratified the Colorado River Compact in 1944 and entered into an agreement with the Department of the Interior for the state's share of the water, subject to various conditions. However, to complicate the situation further, the United States and Mexico signed the Mexican Treaty Obligation in that same year, whereby Mexico is guaranteed 1.5 million acre-feet in years in which there is a surplus. The water is to come equally from the upper and lower basin areas. This is an international treaty and must be observed, even at the expense of agreements among the states. To get some of Arizona's Colorado River water to central Arizona, the state proposed the Central Arizona Project in 1947, but it was to be more than 20 years before the project was approved by Congress.

The first dam built on the Colorado River in Arizona was a small one, Laguna Dam, built a few miles upstream from Yuma (Map 6.9). It is a diversion dam that raises the water a few feet and puts it into the Imperial Canal on the California side of the river. With the use of a siphon, some of the water is brought from the California side to Arizona.

To control the river, a very large dam was needed, and it was the Boulder Canyon Project that brought results (Map 6.9). Hoover Dam, completed in 1936, was built between Arizona and Nevada in Black Canyon, a few miles below Boulder Canyon. It is one of the largest dams in the world and can hold 30.5 million acre-feet. It was not designed to provide irrigation water for Arizona, but it does store water for downstream use, and it is also a major producer of electricity. Most of the power goes to California, and its sale will pay for the dam. With completion of Hoover Dam, Laguna Dam was too small, and another dam was built. This is Imperial Dam, 18 mi. (29 km.) above Yuma, and it diverts water to the All American Canal, with some of the water going into the Gila Canal on the Arizona side and some of it diverted to Mexico.

Two other fairly large dams were built below Hoover Dam: Parker Dam, completed in 1939, and Davis Dam, completed in 1949. Parker Dam, just below the confluence of the Bill Williams River, mostly serves the needs of southern California cities. It was built in conjunction with an agreement between the Bureau of Reclamation and the Metropolitan Water District of Southern California. From Lake Havasu behind the dam, water is pumped through the Colorado River Aqueduct 242 mi. (389 km.) to Los Angeles and San Diego. Power is produced at this dam; half of it is used to pump the water westward, and the other half is sold to help pay for the facilities. Davis Dam, between the Hoover and Parker dams, was built primarily to provide power and to store water for use in Mexico.

The last large dam on the Colorado River in Arizona is Glen Canyon Dam, which was built by the upper basin states and is above Lee's Ferry. It was completed in 1964 and is capable of storing 28 million acre-feet, about as much as Hoover Dam though it has yet to reach capacity. It was planned and built by the Colorado River Storage Project, in conjunction with three other upstream dams, in order to control the upper half of the Colorado

River. It was built primarily to store water to ensure that the water flow obligations to the lower basin states can be met. The dam is also a major producer of power, but its power is for the benefit of the upper basin states. This dam is of particular importance to Arizona because, without it, the town of Page would not exist. Page was established initially as a construction camp after the area had been acquired by the government in a land exchange with the Navajo. The federal government designed the camp as a permanent townsite and spent a great deal of money in building it. Today, though small, Page is significant in that it is the only town within a 75 mi. (120.7 km.) radius.

The Central Arizona Project (C.A.P.), which will take Colorado River water from Lake Havasu to central Arizona, is not scheduled to be completed until 1987. The water will have to be lifted well over 2,000 ft. (610 m.) by about 10 pumping plants and carried almost to the Mexican border to a lake created by a new dam, Charleston Dam, on the San Pedro River (Map 6.10). Two hydroelectric dams on the Colorado were proposed, in Marble and Bridge canyons, but both have been replaced by the Central Arizona Project's investment in the coal-fired Navajo Power Station at Page. Thus the power to transport the water will be available but it will be expensive.

The C.A.P. is extremely controversial. Many are for it simply because it

PROPOSED CENTRAL ARIZONA PROJECT

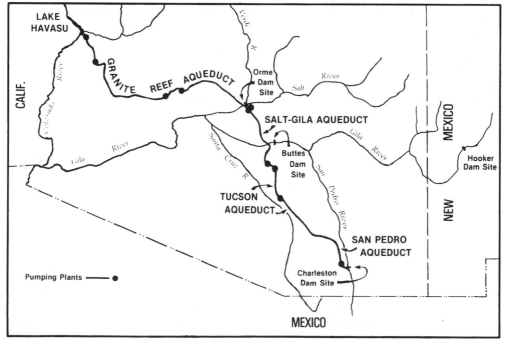

MAP 6.10

will bring water to Arizona, but others consider it to be a "pork barrel–type" of project. Several large dams would have to be built, and one of those, Orme Dam, would have flooded much of the Fort McDowell Indian Reservation. Plans were to fill this lake all winter, but by the end of the summer, it would have been a large mud flat, of little use for recreation. Plans for that dam have largely been abandoned, though there is still some controversy over the dam site. In fact, construction of the C.A.P. south of the Salt River is now in jeopardy.

Another controversial aspect of the project concerns the use of the water. No new land will be put into agriculture, so farmers receiving Central Arizona Project water could pump that much less groundwater. But that would still not save the groundwater level because overdrafts would continue, particularly by farms not receiving C.A.P. water, and the inevitable would only be delayed. Environmentalists and economists argue that if land is to be used for agriculture, then move the farmers to the Colorado River Area where they could use gravity water, thus saving construction and power costs. Such a move would also largely solve the groundwater overdraft problem, for farmers are now doing about 90 percent of the pumping. However, only a change in the law, and not the C.A.P., can end the overdraft on groundwater. There are also arguments as to who would get the water. Cities and irrigation districts are out to get as much as possible, but there will not be enough to go around. The Indians are also demanding more and more water, and since the federal government will pay for it, they want as much as possible.

Another controversial aspect is that the lakes and canals would lose 100,000 acre-feet to evaporation. As an example, perhaps as much as half of the water stored behind the Charleston Dam would be lost to evaporation. The most embarrassing aspect is that there may be no water to put into the C.A.P. canal at all. In order to get the project approved, Arizona agreed that California has first rights to Colorado River water. There should be water for the C.A.P. until the year 2000, but by that date the upper basin states will probably be using most of their allotted water, and there will be no surplus water to put into the C.A.P. canals. All in all, it is a very controversial and expensive project.

Many conflicts over Colorado River water will probably soon erupt, and they will last for years. The upper basin states are now arguing that the flows of the Gila and Salt rivers belong to the lower basin states and that those flows should be deducted from the total amount of water due Arizona. Arizonans are horrified by the thought, but the issue may be taken to court. Also, should the Navajo decide to take Colorado River water for irrigation, they could almost drain the river dry as it passes their reservation. The Metropolitan Water District of Southern California will be gravely affected after the Arizona and Nevada projects are completed in the 1980s, and it is now searching for other water sources. The problem of who gets the Colorado River water is far from settled.

Source and Use of Water in Arizona Today

The majority of water used in Arizona today is groundwater, and unfortunately, the removal far exceeds the natural recharge (Figure 6.6). A great deal of surface water is also used, and 900,000 acre-feet, or 39 percent (18.7 percent of the total water supply), come from the Colorado River. One of Arizona's major goals is to increase its use of Colorado River water, as allowed by the Colorado River Compact, but as mentioned, it is doubtful if the total amount appropriated can increase dramatically over the long term. Well over 90 percent of the water used in Arizona is used in the southern counties (Maricopa, Pinal, Pima, Cochise, and Yuma), and Maricopa County is far and away the major user.

The biggest user of water in the state is agriculture. Included in this category is the livestock industry, but it uses no more than 1 percent of the total. Almost all of the farmed land is in the desert portions of the state and must be irrigated. The total amount of irrigation water needed depends upon the crop (Figure 6.11) and the irrigation technique used, but the average field receives about 3.75 acre-feet of water per year.

FIGURE 6.11

VARIOUS CROPS AND THEIR WATER NEEDS IN ARIZONA

Crop	Water Consumption (in inches)
Cotton	41.2
Safflower	45.4
Soybeans	22.2
Sugar Beets	42.8
Alfalfa	74.3
Barley	25.3
Grain Sorghum	25.4
Wheat	22.9
Grapefruit	47.9
Grapes	19.6
Navel Oranges	39.1
Broccoli	19.7
Cantaloupe	19.1
Carrots	16.6
Cauliflower	18.6
Lettuce	8.5
Potatoes	24.3
Sweet Corn	19.6
Guar	23.1

Source: Consumptive Use of Water by Crops in Arizona, Agricultural Experiment Station, University of Arizona, Technical Bulletin 169, 1976.

Agriculture now consumes 4,294,000 acre-feet each year, which is over 89 percent of all water used (Figure 6.12). Many people argue that entirely too much water is devoted to agriculture, since so much of it is groundwater, since mostly nonfood crops are grown, and especially since agriculture produces only 3 percent of Arizona's total personal income. However, agriculture will continue to be a very heavy user of water until the groundwater sources are exhausted or until water becomes too expensive.

The second largest consumer of water is the municipal and industrial sector (Figure 6.12), and this sector uses 329,000 acre-feet yearly, which is 6.9 percent of the total. This category includes the water used by the cities to irrigate parks, school grounds, and the like, as well as the water used by the citizens. There is a great amount of variation between cities in the amount of water used. The amount of water used per capita in Tucson is considerably less than the per capita use in Phoenix, where the water is much cheaper and more residents try to maintain lawns. Nevertheless, the direct use of water by humans in the state is about 3 percent of the total. Industrial use of water is not great, probably less than 1 percent of the total, and most of what is used is for cooling purposes.

The third major user of water is the mining industry, which uses 131,000 acre-feet, 2.7 percent of the total. This water is used for milling and for concentrating low-grade ores. The water is used wisely, and this industry pays

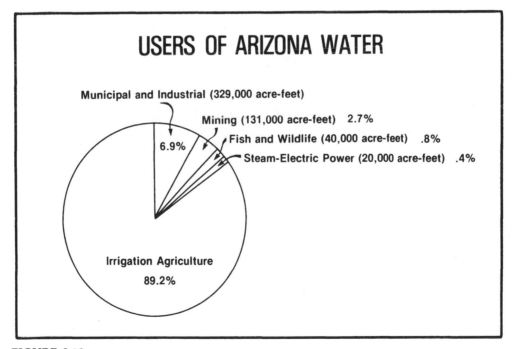

FIGURE 6.12
Source: Arizona Water Commission, *Arizona State Water Plan, Alternative Futures, Phase II* (1977).

high wages. The value produced is about equal to that of agriculture, but mining uses much less water.

The last two categories use a relatively insignificant amount of water. The water used for fish and wildlife maintenance amounts to 40,000 acre-feet each year, or less than 1 percent of the total, and most of this use is to maintain wildlife refuges along the Colorado River in Mohave and Yuma counties, and a little bit is sewage affluent from Phoenix that is used to help maintain a riparian habitat in the Gila River below the city. A final category of user is the steam-generated power industry. This industry needs quite a bit of water for cooling purposes, but compared to the overall water use in Arizona, it uses a small amount. This industry currently uses 20,000 acre-feet each year, or 0.4 percent of the water used in the state.

Water use is a complex and sensitive issue in Arizona. Most of the water is used by agriculture, and unfortunately, it is not used wisely. Most of the water used in the state is groundwater, but 88 percent of all pumping is overdraft, or the mining of water. The water will, therefore, be exhausted someday, and this pumping will end. The state plans to bring water to central Arizona from the Colorado River, but that action will not solve the problem. Only after there is a positive change in the groundwater laws will water be used wisely – and groundwater saved for future generations.

7

ARIZONA'S EVOLVING ECONOMIC BASE

Arizona's economy is a varied one. Agriculture, ranching, lumbering, and mining have traditionally been important, and they are still significant to the economy, but their relative importance is declining. Today manufacturing is the most important industry in Arizona, and it far outstrips other sectors of the economy in significance. Most of the manufacturing is in electronics, so it is non-polluting, high paying, and very desirable. Tourism is also important in the Arizona economy, because Arizona has many natural wonders that attract visitors from around the world. There is rapid growth and the state's economy is expanding, and therefore the construction industry is also dynamic and growing. Arizona thus has a viable and dynamic economy, an economy that is slowly changing and evolving to fit the needs of the latter half of the twentieth century.

AGRICULTURE

Arizona is a large state, but only 1.9 percent (1,327,670 acres [537,706 ha.]) is agriculturally productive, and well over 90 percent of the agriculture is found in the southern counties (Map 7.1). Maricopa County is the largest agricultural county, and it ranks as the fifth leading agricultural county in the nation, with over 500,000 acres (202,500 ha.) under cultivation. Other important agricultural counties are Yuma, Pinal, Cochise, Graham, and

AGRICULTURAL LAND - 1974

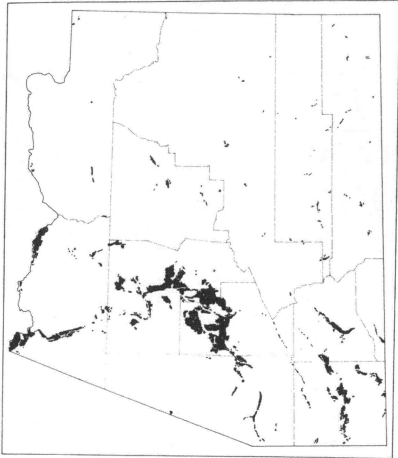

MAP 7.1
Source: Office of Economic Planning and Development, "Existing Arizona Land Use," 1978.

Pima. There are also small, isolated patches of agriculture in northern Arizona in various floodplains and mountain valleys. On the Hopi and Navajo reservations, corn for local consumption is the dominant crop, and it is grown in scattered locations. The total value of Arizona's crops in 1977 was over $689 million.

Although urban areas are spreading out over the agricultural land, the total area under cultivation has increased rather than decreased; in the decade between 1967 and 1977, over 100,000 additional acres (40,500 ha.) were put into agriculture. Perhaps urbanization can help save water, since an acre of houses uses far less water than an acre of cotton. If the inhabitants of the houses work in manufacturing or service industries, Arizona's economy would be much stronger, because the taxes on homes and industry are much greater than on land that is used for agricultural purposes. Urban users can also afford to pay considerably more for the water they use than the farmers can. Thus, the conversion of farmland to urban uses seems to be another possible solution to the water problem.

The typical farm in Arizona is not a family operation. Arizona farms are distinguished by their large size, mechanization, high capitalization, substantial use of seasonal labor, double-cropping, and professional management. It is truly agribusiness. High per acre yields are also common, often double the U.S. average. There are a few independent farmers, who are usually located in older agricultural areas, and most of them inherited their land. All agricultural land is very expensive, and it is just as expensive to bring raw land into production. The cost is far beyond the reach of most individuals, therefore, corporations and agribusinesses are dominating this industry more and more.

Irrigation is another characteristic of farms in Arizona, and they are the greatest users of Arizona's most precious and critically limited resource, water. Much of the water used for farming is groundwater that is rapidly being depleted. Water can be very expensive – when a farmer is pumping from a very deep well – or it can be very cheap – as it is in the bounds of the Salt River Project. The cost of water will often determine what crop is grown; for instance alfalfa is grown in the Phoenix area where water is cheaper, and tree crops are grown in areas where water is too expensive for cotton.

Most irrigation is surface irrigation, also known as floodwater farming, and it is simply the flooding of the fields or rows (Figure 7.1). This is an old but a very wasteful method. Drip or trickle irrigation, whereby water drips from an underground pipe, is far and away the most efficient method of irrigation, and it is usually used to irrigate tree crops. Few Arizona farmers use this technique, but in the future, as water costs rise and this method is made more efficient, its use will spread. Sprinkle irrigation is also far more efficient than flooding, but there are costs associated with putting the water under pressure and with all the hardware needed. Most farmers, unfortunately, find it cheaper to waste water with surface irrigation rather than to adopt the newer techniques. But a variation of the traditional surface irrigation technique does show great promise for conserving water and energy. This is level-basin irrigation, and it involves a careful leveling of the land, with laser-controlled grading equipment, and the building of dikes around the fields. In level-basin irrigation only a little water is put on the fields, all through gravity flow, and there is no loss of tailwater on the lower end of a field. Water efficiency is between 70 and 90 percent, equal to the efficiencies of sprinkler systems. Level-basin irrigation can't be used everywhere, and it is expensive to initiate, but its use will spread widely.

Croplands are found mostly in the floodplains extending along the rivers,

FIGURE 7.1. A field being irrigated by syphons west of Phoenix.

and such areas are common in southern Arizona. These plains have gently sloping surfaces that are largely naturally graded and, thus, are excellent for irrigation. In such areas there are also streams, and relatively shallow groundwater bodies. The soil in such areas is usually excellent, except for perhaps a low content of organic matter, and is formed of recently deposited alluvium. Although the soil is rich, the farmers in Arizona are heavy users of fertilizers, especially nitrogen and phosphorus.

The farmers are also heavy users of insecticides. Arizona was largely free of insects when the whites arrived; it had a hot, dry climate with few host plants. By the early twentieth century, because of irrigation and insect introductions, there were many pests, and the entire area had become a giant hothouse and insect incubator. The spraying of insecticides is now the rule, but such spraying by aircraft near the residential areas poses a serious and growing problem (Figure 7.2). The state maintains agricultural inspection stations along the roads leading into the state, with the hope of keeping out pests such as spider mites from California and fire ants from the South, but in an economy move, these inspection stations may be closed. The only insect of real use to the farmers is the honeybee, which was introduced from California in 1872. The bees are critical in the pollination of watermelons, cantaloupes, tangerines, vegetable and alfalfa seeds, and several other

FIGURE 7.2. Spraying a cotton field from the air.

crops, and the farmers will pay beekeepers to put hives near their fields.

It can be argued that agriculture is very important to Arizona's economy, or it can be argued that it is of little significance and only a great waster of water. Agriculture is a stable industry. It is also an important basic sector of the economy, and in terms of the value of its products, agriculture ranks third behind manufacturing and mining. Conversely, agriculture employs only 2.3 percent of the labor force (22,800, of which 16,000 are hired workers and the rest are owners or unpaid family members) and only produces 2 percent of the state's personal income ($300 million in 1977). The processing of agricultural goods is classified under manufacturing, which affects the figures somewhat. A few agricultural workers are well paid, but most are stoop laborers who are living in deep poverty. Agriculture, therefore, does not provide a good living for those involved in it. It may be argued that the food produced by Arizona's farmers is needed, but such is generally not the case. In the early days, there was a dependence on the locally grown foods, but beginning in the early 1900s, there was a shift in agriculture to growing cash crops for export outside the state and importing food for consumption from the humid East. Highly developed technology and efficient transportation systems make this possible and Arizonans can be greatly isolated from their harsh environment. Since the farmers do not grow crops to feed the local populations (except for some citrus and vegetables and, indirectly, meat and milk), and since farmers use almost 90 percent of the water, perhaps agriculture is not indispensable. Water is the most critical and limiting resource in a desert, and it can be argued that it makes little sense to use the water for agriculture, especially the part that comes from the overdraft of groundwater, a situation that, sooner or later, will have to end anyway.

Cotton is traditionally the most important crop. There are basically two types of cotton, short staple and long staple, and both are grown in Arizona. Short staple cotton, sometimes called upland cotton, is native to the South, and it has short fibers, each about an inch long. Since upland cotton produces a short fiber, it brings a lower price but it is a good producer (in 1977 upland cotton produced 979 lb. per acre [1,098 kg./ha.] statewide, and yields up to 1,500 lb. per acre [1,683 kg./ha.] are not uncommon). In contrast, long staple cotton, often called Egyptian cotton, has long individual fibers, about 2 in. (5.08 cm.) in length. Long staple cotton brings a higher price because it produces a superior product, but yields are lower (694 lb. per acre [778.5 kg./ha.] in 1977). Long staple cotton is suited to desert conditions and cannot be grown in the humid East.

Several varieties of long staple cotton have been grown in Arizona; the first variety grown was the Yuma variety, then the Pima and others, and the Pima S5 variety is grown today. The two types look different and can be easily recognized (Figures 7.3 and 7.4); the long staple is quite tall, often over 7 ft. (2.13 m.), and the short staple is a smaller plant, often under 3 ft. (.9 m.). Regardless of which type is grown, all of Arizona's cotton is top quality, since there is little rain to spoil it.

Cotton has a long history in Arizona. The Indians grew cotton long before the arrival of the whites, but it was different from that grown today. Also, many of the first Americans in the state were from the South and Texas, and they knew cotton well and experimented with it. They quickly realized that cotton grew well in the area, but there was no great demand for the product, and there were no local cotton gins. The cotton industry began in Arizona in 1912 when 400 acres (162 ha.) of the Yuma variety of long staple cotton were harvested (Figure 7.5). By 1917, there were several cotton gins in the state, and cotton had become Arizona's most important crop.

The reason for this boom in the cotton industry was that there was a great demand for long staple cotton to be used in making automobile and truck tires and airplane fabric. These products were in great demand about the beginning of World War I, and the long staple cotton had to be grown in the arid West. Big industry soon arrived. By 1918 Goodyear Tire and Rubber Company controlled 75,000 acres (30,375 ha.) in the Salt River Valley and another 17,000 acres (6,885 ha.) in the Yuma area, most of it designed for cotton production. By 1920, Arizona had 230,000 acres (93,150 ha.) in cotton.

For every boom in Arizona history, there seems to be a bust, and the collapse of the cotton industry came in 1921. The reason for the collapse was a drop in the price of cotton, in that year the price fell from 34 cents to 15 cents a pound. Goodyear began to dispose of its land, though it still owns some large tracts, and the small farmers turned to other crops, such as dairy-

FIGURE 7.3. A man standing next to a field of long staple cotton. Note the height of the cotton plant.

FIGURE 7.4. A field of short staple cotton. The plants have been defoliated prior to picking, and it looks like a sea of white.

COTTON ACREAGE IN ARIZONA, 1912-1978

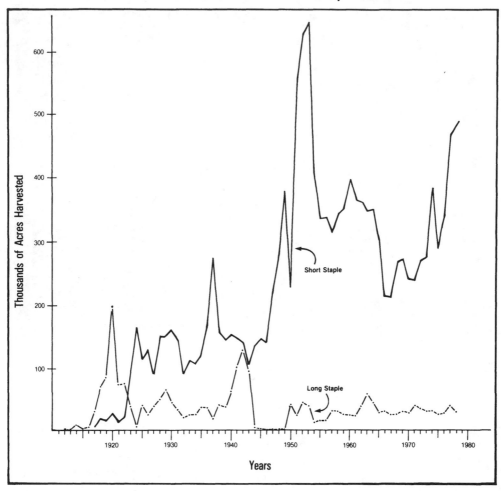

FIGURE 7.5
Source: Arizona Crop and Livestock Reporting Service, *Arizona Agricultural Statistics*, various years.

ing, grains, and vegetables. Cotton quickly made a comeback, however, and since then production figures have fluctuated greatly in different years, mostly because of economic conditions, not because of environmental factors (Figure 7.5).

Cotton also affected Arizona's population, because it used to be a labor-intensive crop. In the early years, all cotton had to be picked by hand, and the only groups to have a tradition of picking cotton were the whites and blacks from the South. The Mexicans did not have that tradition, and few of that group were involved in the industry in Arizona. Many southerners were recruited to work in Arizona's cotton fields, particularly between the depression years of the 1930s and the late 1940s (Figure 4.29). Many of them became permanent residents, others were only migrants, but during a cotton harvest thousands could be found in the cotton towns of south-central Arizona. Mechanization of the cotton industry began in the 1940s, and by the late 1950s little cotton was picked by hand, and the former cotton pickers had drifted into other occupations. The cotton plant was changed to fit the cotton picking machine, and now it cannot be picked by hand because the cotton does not easily leave the boll (Figure 7.6). Also developed was a machine called a "rood" to pick the cotton that had fallen on the ground (Figure 7.7). Cotton picked by a rood brings a poor price, but at least it is not wasted.

Cotton acreages vary greatly from year to year, but today the crop dominates farming in Arizona. In 1977, 556,000 acres (225,180 ha.) were in cotton. Although the Arizona cotton industry began with long staple cotton, only about 7 percent, or 41,500 acres (16,807 ha.), are in this type of cotton today, and 93 percent is in short staple. Cottonseed has also become a very important product, but unfortunately a mold that produces aflatoxin is sometimes found in cotton produced below 1,800 ft. (549 m.). Such infected cottonseed is sometimes fed to animals, but it can kill them, and perhaps it also indirectly affects the humans who eat the animal products, because aflatoxin is reportedly cancer causing.

In 1976 cotton sold for 71 cents per pound, so cotton acreage increased in 1977, but in that year cotton sold in the range of only 51 cents per pound. Farmers are sensitive to such drops and immediately reduce their cotton acreage. Cotton also consumes a great deal of water, and in the future, as the cost of water climbs, farmers will turn to other crops. However farmers are hesitant to make this shift, because cotton has a potential for profit, it is a known crop, and many farmers have invested in cotton pickers that cannot be used for other crops.

Another important crop in the state is alfalfa, and it has been important since territorial days. It was one of the first crops grown in the area by the Americans, and the economy of the Salt River Valley in the early era was based on the production of alfalfa for sale to the military for use as fodder. Since then, it has remained an important crop in the state, and about 200,000 acres (81,000 ha.) have been in alfalfa in the last few years. Alfalfa fields are grazed by cattle and sheep, and alfalfa is cut for hay, pelletized, or used as green chopped feed. Most of the alfalfa is used locally, but some is shipped out of the state, particularly to California. Alfalfa produces very high yields; it can be cut up to eight times a year and averages 6.5 tons of hay per acre (14.6 t./ha.). The only real problem with alfalfa is that it consumes 6 acre-feet of water per acre per year, which means that it can be grown only where water is plentiful and cheap, such as in the Salt River Valley.

Various grains are important food crops. These crops also use very little water, about 2 acre-feet per acre per year, and are very productive, produc-

FIGURE 7.6. A cotton picker at work. It is moving toward the camera, picking two rows at a time.

FIGURE 7.7. A rood at work picking up cotton from the ground. Courtesy, Ray Henkel.

ing well over 2 tons of grain per acre (4.5 t./ha.) each year, so these crops probably will remain important in Arizona's agriculture. Barley is one of the important grains, and in the past few years an average of 60,000 acres (24,300
ha.) each year has been planted in barley. Wheat is the most important grain
crop, and in 1977, 140,000 acres (56,700 ha.) were planted in this crop (Figure
7.8). Two important types are grown, winter wheat and durum. In 1975, there
were 40,000 acres (16,200 ha.) in durum wheat, but that figure jumped to
319,000 acres (129,195 ha.) in 1976. The reason for the jump was that
foreigners, especially the Italians, were buying it for pasta products. The price
fell as quickly as the boom occurred, and in 1977 only 85,000 acres (34,425
ha.) were in durum wheat. Sorghum, an African native, is another important
grain crop, which is mainly used for cattle feed. It is easily recognized, since
the leaves resemble corn, and the grain is produced on stalks rising about the
short plant (Figure 7.9). Corn is another significant grain crop. Formerly
fewer than 20,000 acres (8,100 ha.) were devoted to corn, but recently corn
production has risen. In spite of the fact that corn has been grown in Arizona
since the days of the early Indians, the climate is not ideal for its production,
particularly the chilly desert nights. Some sweet corn is grown, but most
of the corn crop is chopped up—leaves, stalk, cob, and all—and fed to cattle (Figure 7.10). Oats is another grain crop grown, but it is of little significance.

Root crops are also of significance in the state. Potatoes have been a traditional crop in Maricopa County since 1929, such as around Queen Creek, but
the potato acreage is declining, and today only about 6,000 acres (2,430 ha.)
are planted. Most of the potato crop is sold to potato chip companies. Sugar
beets are another important root crop, and they are also grown in southern
Arizona. There was an interest in sugar beets in the early twentieth

FIGURE 7.8. A field of wheat. Beyond the wheat is an old date orchard.

FIGURE 7.9. A field of sorghum. Note how a few of the plants grow tall. This is not desirable; farmers would prefer that the plant grow close to the ground and spend its energy producing grain.

FIGURE 7.10. Corn being ground up – stalk, leaves, cob, and all – in order to feed it green to cattle. Courtesy, Ray Henkel.

century, but a factory producing sugar beets in Glendale went bankrupt in 1913. Beginning in 1936, sugar beet seeds have been produced in Arizona for use in the American Northwest, and in 1965 the Spreckels Sugar Company built a factory in Chandler. During the harvest season, trucks and trains loaded with these large tubers can be seen moving to this factory. These plants are growing in rotation with other crops, and anywhere from 10,000 to 20,000 acres (4,050 to 8,100 ha.) are in sugar beets. There is some question, however, whether or not the sugar beet industry can survive in Arizona, because of the pests and plant diseases that affect the plants.

Citrus crops are another major type of crop in the state, grown in the Phoenix and Yuma areas, and they were introduced into the Salt River Valley from California in 1870. These crops, particularly the lemons, are easily damaged by freezes, so they are mostly grown on slopes where there is air drainage; the colder air moves downslope to the lower elevations, and the higher sites remain a bit warmer. Large windmill-like structures are often seen in the citrus groves, and they are used to circulate the air on colder nights. The major citrus crops are Valencia and Navel oranges, which mature in the winter and spring, respectively; lemons, grown mostly in Yuma County; grapefruit, grown mostly in Maricopa County; tangerines; and various other kinds of fruit. A little less than 60,000 acres (24,300 ha.) are planted in citrus in Arizona. It is all picked by hand (Figure 7.11). Ap-

FIGURE 7.11. Young Mexican Americans picking citrus. All citrus must be picked by hand.

proximately one-third of the crops is sold fresh throughout the United States, and the rest is exported or processed. The foreign customers are Taiwan, Canada, New Zealand, the Common Market countries, and many others.

Another significant crop is grapes. There are only a little over 3,000 acres (1,215 ha.) in grapes, but they have been commercially grown since the 1880s, and there has been a constant interest in this crop. Almost all grape production is in Maricopa and Yuma counties, and well over half the acreage is the Thompson Seedless variety, which is ideal for raisins. Most of the grapes are now grown by large corporations, because it is an expensive business to enter, each acre requires about a $2,000 investment, and there is about a four-year wait before there is any return on an investment. It is also a very risky venture. Arizona grapes are sold fresh, as raisins, or shipped west to be mixed with California grapes to make inexpensive wine. Prospects for the local grape industry look promising (Figure 7.12). There is increasing interest in redeveloping a wine industry in the state, and the plant uses very little water, each acre of vineyard uses only about 1.5 acre-feet each year. The major problem for the grape industry in Arizona is that the summer rains occur just as the grapes are about to mature. Such a rain will cause the grapes to mildew and rot, and often an entire crop can be lost. Most new grape orchards are being planted in areas the farmers think are relatively safe from the summer rains.

FIGURE 7.12. A large, young grape orchard west of Phoenix.

Date growing has never been truly successful. It probably has had a greater impact on tourism than on the agricultural industry, because although a date orchard is a sight tourists remember and appreciate, it seldom produces a crop that is profitable. At least 150 varieties have been tested, and the Deglet Noor became the leading commercial variety in the United States. Southern Arizona, however, has a very wet August, and this causes problems within the date cluster, particularly with the Deglet Noor. The major varieties in Arizona are Khadrawy, Zahidi, Halawy, Kustawi, and Medjool; the Deglet Noor is limited to the drier Yuma area. It is common to see bags around the fruit in central Arizona; these keep out birds and, more importantly, keep the fruit dry. The Salton Sea area of California has a hotter and much drier summer, so dates thrive there. Date production is a very labor-intensive crop, since laborers must climb the trees several times a year, so date production today is a small and fitful busines in Arizona.

There are also many tree crops in the state. These include pecans in particular, but also almonds, apples, apricots, nectarines, peaches, and plums. Tree crops use less water than cotton, and in areas where water is getting expensive, pecans are being grown as an alternative crop. The big problem with such crops is the long wait before there is a harvest, and in the meantime a blight or collapse of the market could mean disaster.

The vegetable and melon industries are very important to the state, with about 60,000 acres (24,300 ha.) in production, annually worth about $100 million. Arizona today ranks fourth in the nation—after California, Florida, and Texas—in the production of these products. Many Mexican Americans are employed in these fields, since vegetables require much stoop labor, so vegetables are very important in that they provide labor and income for residents of the state. Another advantage to vegetables is that they require very little water; a crop of lettuce, for example, needs only one acre-foot. There is also a great deal of double-cropping of vegetables. Most are shipped out of the state to urban markets, particularly to the West Coast and the Midwest. These fresh vegetables arrive at the urban markets very early in the season, and, thus, they bring high prices.

Lettuce is far and away the most important vegetable crop. Well over half of the total value and total acreage of the vegetables in Arizona is in lettuce, and most of the production is in Yuma County. This crop is also particularly important to the laborers, for the lettuce season is the one time of year when they can earn a decent income (Figure 7.13). Lettuce is also double-cropped and has several harvesttimes in the state, which helps the laborers. As with other crops, there have been many changes in the lettuce industry. The types now grown produce 150 percent more than those grown in the 1920s; they are also better able to withstand transportation; and new techniques, such as vacuum cooling, allow them to arrive fresh at the markets. The second crop in acreage and value is cantaloupes, and again Yuma County is the major producing region. Other important crops are dry onions, watermelons, carrots, cauliflower, honeydew melons, and broccoli, and

FIGURE 7.13. Migrant laborers harvesting a field of lettuce in the early 1960s.

there are many others of lesser significance.

Seed production is important to Arizona. The major advantages Arizona has over other areas are that the climate is dry, there is little chance for rain as the seeds mature, and the fields can be isolated. Seeds can be grown in isolated desert areas under very controlled conditions; water can be put on or removed at exact times, and there are no strange pollens entering the fields to contaminate the seed crops. Bermuda grass seeds are particularly important, about 90 percent of the nation's production comes from the Yuma area, and Arizona produces about half of the nation's supply of sugar beet seed. Also significant is the seed production for alfalfa, sorghum, and a few other crops.

Another industry closely related to agriculture is landscaping. Because of a concern today for aesthetics and image, a great deal of money is spent on landscaping homes, schools, businesses, and the like. It is a very lucrative business, and it actually produces more wealth than the vegetable industry.

LIVESTOCK INDUSTRY

When people in the East think of Arizona, they usually think of ranches and cowboys, but that is a false image. Arizona has a desert environment, which can support only a few cattle, and among the fifty states Arizona ranks far down the list in total cattle production. Cattle raising is much more important in other states, particularly in the midwestern and southern states. Traditionally, weather conditions, particularly droughts or wet periods, determined how many cattle were raised in the state, but since the growth of feedlot operations, it has been economic conditions and price that have largely determined such totals.

Development of Ranching

Ranching has been an important part of the Arizona landscape since the arrival of the Spaniards. Spanish land grants were awarded in southern Arizona, and by the early nineteenth century ranching was a big industry. It is hard to believe that the ranches really earned much money, however, because there was no local market for the cattle. With Mexican independence many changes began to occur, and the ranching industry declined.

When the Americans arrived in Arizona the ranching industry had been destroyed. When Captain Cooke led the Mormon Battalion through the San Pedro Valley only a few wild cattle remained from a once thriving industry. The only Mexicans remaining were a few along the Santa Cruz River, but there, too, ranching had ended. It was the Apache problem that had broken the ranching industry. The Spanish government had bribed the Indians, and there had been peace, but soon after Mexican independence, the Apache again went on the warpath. By the beginning of the American period, mounted Apache were raiding far into Mexico, and one of their chief aims was to acquire cattle.

The Americans were familiar with ranching techniques long before they moved onto the Great Plains. Cattle had been brought over by the earliest colonists, and cattle raising was a well-established industry. Southerners, mostly poorer white people of English and Scotch-Irish descent, raised large herds in the mountain forests by simply allowing the cattle to roam free. They annually rounded up the cattle, branded them, and sold them to plantation owners as food for their slaves. Many southerners later went into Texas to begin ranching after the cotton lands to the east were exhausted. Ranching would probably have evolved even if there had been no Spanish influence, as it did in Australia and Southern Africa. The Spanish, however, made many contributions to American ranching, and it was in Texas that Spanish and American ranching techniques met and blended to form the early and romantic cattle industry.

The Spanish-Mexican contributions to the cattle industry are varied. Roping techniques were new to the Americans, but the Mexicans were real masters of the rope, and it was an idea that the Americans readily accepted. Going along with the rope were the pommel and the *tarabi,* a tool used in making ropes of horsehair. The Mexicans also introduced spurs, chaps, and other items to the Americans, plus many words. The Americans changed some of the techniques to fit their own ideas, and what happened to the rope is a good example. The Mexicans used horsehair and braided leather to make ropes that were very thin, about the size of a man's little finger. After roping an animal, the Mexicans would play it like an angler plays a fish until it tired; otherwise the animal would have broken the thin rope. The Americans like brute strength, so they developed very strong ropes that would not break when a large animal managed to get a good jerk on one.

Some ranching operations remained different, however, such as the brands and branding irons used. The Mexican brands had little rhyme or reason to them, and the Americans derisively referred to them as "chicken scratchings." The Americans liked clear, simple brands, based mostly on letters of the alphabet, and ones that could be easily read, from left to right or top to bottom. This American system had the great advantage that brands could be quickly and easily recorded and checked because they were chiefly based on the alphabet. The branding irons were also different, the Mexicans used a short handle and the Americans a long one. The Americans would make fun of the Mexicans for getting on their knees to brand the cattle, but the Mexicans would respond that they could put the brand right where they wanted it; there were advantages to both ways. The blending of the techniques led to the early western ranching industry, and from Texas the ideas spread northward and westward.

Cattle ranching did not quickly develop in Arizona in the early American period. There were a few wild cattle, but there was little demand for them. In contrast, vast herds had been abandoned in Texas, and there was a good market for them in the nearby southern states. Large numbers of Texas cattle were also driven across Arizona to the California goldfields, because cattle brought $300 a head there in 1849, but in Texas the cattle could be had for about $4 each. A little ranching developed in Arizona when the California market collapsed about 1853, and the Arizona range was stocked with Texas longhorns or Sonoran cattle of equally poor quality. However, the Indian troubles during the Civil War quickly ruined this industry.

The ranching industry started up again after the Civil War, and again, its center was in Texas. Vast herds had developed there during the war, and once again they were there for the taking. They commanded high prices in the North, and entrepreneurs quickly began moving them to that market. They were first driven to the Mississippi River for transport northward on steamboats. Later, when the railroads began to enter the Great Plains, large herds were driven to railheads such as Abilene and Dodge City. Meanwhile, there were only a few cattle in Arizona, and there was no way to easily get them to a large market. Ranching in Arizona had to await the coming of the industrial revolution.

The industrial revolution allowed the full development of ranching as we know it. New firearms, for example, allowed a coupling of firearms and the horse. A Kentucky rifle was really too awkward to use on horseback, but the Colt revolver and the Sharps and Winchester rifles—with their paper

FIGURE 7.14. A windmill pumping water into a concrete tank in southeastern Arizona.

and, later, metallic cartridges—allowed an armed man on horseback to be a formidable opponent. The industrial revolution aided in many other and more positive ways. Transportation in the West was always a problem until the problem was solved by the railroad. The railroads made the long-distance transportation of cattle possible, and they brought in lumber, salt, windmills, and all the other things needed by the ranchers.

Water was always a problem in the West until the invention of the windmill (Figure 7.14). Cows will usually go no farther than 3 mi. (4.8 km.) from water, and they tend to "eat up" the range near a waterhole and "ignore" the good grass of the more-distant areas. The early homesteads were along the streams, and the ranchers allowed as much grazing of the surrounding land as was practical. The American windmill developed in New England in 1854, and it evolved from the locally used brine pump windmills, which in turn had developed from the windmills of the Low Countries. With windmills scattered across the land, ranchers could effectively exploit the entire range. By the time the windmill began to gain acceptance, techniques for drilling narrow, deep wells had developed, and these could be used with the windmills. The windmill had many good qualities: (1) they were inexpensive, costing about $150 to $160 each; (2) they were easy to erect and lubricate; and (3) they were efficient and cost little or nothing to operate.

Soon all the large ranches had professional "windmillers," and they received substantially higher wages than the ordinary cowboy ($75 per month as compared to $45). Windmills were also important for the railroads, for they were used to pump water into tanks for use in the engines' boilers.

Fencing in the West was another problem that was solved by the industrial revolution. The early settlers had tried various fencing techniques such as (1) wood fences, but they were too expensive; (2) stone fences, used by the Germans in the San Antonio area, but their building was too laborious a task; and (3) hedges of osage orange, cherokee rose, or prickly pear. None of these techniques worked. In the early ranching era, there was simply no fencing, and the cattle grazed on the open range. This procedure caused many problems for the farmers and did not allow for controlled breeding.

The fencing problem was solved with the invention of barbed wire in northern Illinois in 1874. Barbed wire had many advantages: (1) it was inexpensive compared to other kinds of fencing material; (2) it worked, after only one experience with it, animals respected it; (3) it took up little space and made no shade; (4) it was durable; and (5) it stood up to the high winds and caused no snowdrifts. One salesman carried away with his own exuberance is reported to have said that it was "light as air, stronger than whiskey, and cheap as dirt." But it was expensive for the ranchers who had a lot of land to fence. The first extensive use of barbed wire by a cattleman was in 1881 when a Texas rancher fenced 125,000 acres (50,625 ha.) with it at a cost of $40,000, a large sum even by today's standards. Thus, the cattlemen fought fencing and wanted to retain the old open range system. They thought that if the farmers wanted to keep the cattle out of their fields, the farmers should be the ones to fence, and the farmers certainly did fence a great deal in this early era. But in the East, sentiment was with the farmers. The easterners thought the cattle barons were getting rich on free government land, and in the end, it was the ranchers who were forced to fence. The average cowboy was stuck with jobs he detested: cutting fence posts, digging holes, and stretching barbed wire.

The acceptance of barbed wire completely changed ranching. It meant the end of the long cattle drives, and enclosing the range also led to the end of the longhorns, because with the fencing, there could be selective and controlled breeding. Many Hereford bulls were imported, mostly from farms in the Ohio River valley that specialized in Hereford production, and the longhorn bulls were sold for meat. The longhorn, therefore, was not killed off but was bred out of existence. With barbed wire, the old open-range ranching changed to become stock farming.

Thus cattle ranching as we know it developed in Arizona in the early 1890s. The old ranching techniques had been used, but the industrial revolution, and all its contributions, allowed cattle ranching to truly be a success in Arizona.

The Cattle Industry Today

Livestock production is of great economic importance to Arizona, and it is usually about as significant as agriculture, though by the late 1970s it had fallen far behind agriculture in the value of its products. At least 95 percent of the livestock produced are cattle, the rest being goats, sheep, hogs, and poultry. In early 1980, there were 1,050,000 head of cattle in the state. The average price was $465 a head, so the total value of the cattle in Arizona was $488 million. The value of cattle fluctuates greatly through time, so the value changes quickly. The total value of all livestock sold in Arizona in 1977 was over $508 million.

The cattle industry can be divided into two broad categories: intensive production (mostly feedlots and dairying) and extensive production (ranching). In early 1980 there were 505,000 head of cattle in feedlots and dairies in the state and another 545,000 on the range. Therefore, only slightly more than half the cattle are on ranches, and the rest are in the agricultural oases of the state (Figure 7.15). This has been true for the last several years.

The typical ranch in Arizona usually involves only a little privately owned land, often only 160 acres (64.8 ha.); the rest of the land grazed is public

FIGURE 7.15

CATTLE IN ARIZONA, JANUARY 1, 1979 (IN THOUSANDS)

County	Cattle on feed	Milk Cows and Replacements	Range Cattle	Total
Apache	x	x	43.8	43.8
Cochise	x	x	84.2	85.5
Coconino	x	x	68.8	68.8
Gila	x	x	43.8	43.8
Graham	x	x	21.9	27.9
Greenlee	x	x	21.9	21.9
Maricopa	108	79.7	53.1	240.8
Mohave	x	x	56.2	56.2
Navajo	x	x	43.8	44.1
Pima	x	3.1	28.1	31.2
Pinal	258	1.0	31.2	290.2
Santa Cruz	x	x	18.8	18.8
Yavapai	x	x	75.0	75.0
Yuma	117	x	34.4	152.0
Total	490	85	625.0	1,200.0

x Numbers too small to warrant quantitative estimate.

Source: Arizona Crop and Livestock Reporting Service, Arizona Agricultural Statistics, 1978

FIGURE 7.16. A feedlot south of Phoenix. Note the shade provided and the road through the feedlot on which trucks with hoppers will drive, dumping feed into the troughs along the roadside.

land, such as grazing districts, national forests, state lands, and Indian reservations. There are about 1,500 ranches in the state, so the average ranch has approximately 416 head of cattle. Many ranches today are parttime operations and provide only supplementary income, but a few are quite large.

There is grazing in almost all parts of the state, and the only areas to receive no livestock pressure are preempted lands (urban areas, croplands, highways, and certain federal preserves). Some of the lowest elevations, such as the dry desert in the southwestern part of the state, are only grazed in the wintertime, whereas, the highest elevations are used for grazing only in the summertime, and then usually only by steers.

The best ranching land is found between 3,000 and 6,000 ft. (914 and 1,829 m.). The southeastern part of the state is considered the best ranchland (Figure 2.24), because most of it is grassy and can be used year-round, with the cattle moving to higher and lower elevations depending on the season. The same conditions exist in the mountainous central portion of the state, especially in the less-rugged areas, and this is the principal rangeland in Arizona. Cochise County in the southeastern part of the state and Yavapai County in the central region have far and away the greatest number of range cattle. There is also grazing on the Plateau, since much of it is in grass and is excellent for cattle, particularly in the summers. However, Navajo and Hopi reservations are used mostly for sheep rather than cattle.

Almost half of the cattle in the state are in feedlots, not on the range (Figure 7.16). A feedlot is where the cattle are "finished." They are kept in small pens, where they get no exercise, and are fed concentrated foods. Thus, they put on weight, and apparently the tenderness, flavor, digestibility, and nutritional value of the meat are improved. Cattle are generally brought to the feedlot weighing between 400 and 600 lb. (181 and 272 kg.), and they are ready for slaughter when they weigh about 1,000 lb. (454 kg.). Most cattle in the feedlots in Arizona are fed on a custom basis; that is, they are not purchased by the feedlot owners, but the cattle owners pay for the boarding and fattening of their cattle.

The total number of cattle in the feedlots varies. In 1973 and 1974, when the price of meat rose dramatically, there were considerably more cattle in the feedlots than there are now. There is also a seasonal variation. The summers are too hot for the cattle to gain much weight, so the feedlots are fuller in the cooler months.

The fattening of cattle is an old business in Arizona, and cattle were fattened in the alfalfa fields as early as the 1880s. Originally, fattening was only a casual business, and the cattle were fattened in the irrigated pastures until well into the twentieth century. But in the 1950s, it became big business

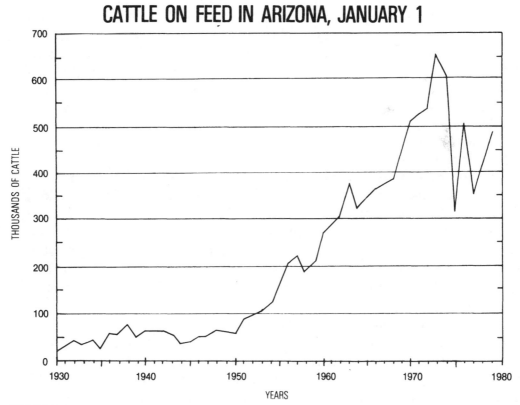

FIGURE 7.17
Source: Arizona Crop and Livestock Reporting Service, *Arizona Agricultural Statistics*, various years.

(Figure 7.17), and the cattle were kept in dry lots rather than in pastures, and the feed was taken to them. The peak of this industry was in the winter of 1973, when well over 600,000 head of cattle were on feed. Since then the number of cattle has fluctuated, depending on economic considerations, and there are anywhere from 320,000 and 500,000 cattle on feed at any one time.

Arizona ranks eighth among the cattle-feeding states. Not enough cattle are produced in the state to supply the feedlots, so cattle have to be shipped in from long distances. Most of the cattle transported into the state come from Texas and the South (Map 7.2). In fact, cattle raised in Arizona do not do well in the local feedlots because of disease. Southern cattle are raised in confined pastures and are exposed to more diseases, so they develop more immunities than the cattle raised on the rangelands in Arizona. Arizonans cannot consume all the meat, and the main market is in California, so Arizona is just a stopover in the shipment of cattle westward (Map 7.3). All of the major meat packing plants are near Phoenix, the major consuming area in Arizona, but not all of the cattle fattened in Arizona can be slaughtered there, so many are shipped live to California.

The typical feedlot in Arizona is enormous. All hold at least 2,000 head of cattle, and the biggest can accommodate up to 80,000. Arizona is a dry state, and that is an advantage in feeding cattle, because a feedlot can get very messy after a series of heavy rains. But Arizona is, in fact, too dry, and since dust stirred up by the cattle can cause lung problems, sprinkler systems are often used around the feedlots to help control the problem. The typical feedlot also has some provision for shade, usually just a ramada-type structure in the middle of pens. Cattle will put on more weight if they have shade, and it is particularly desirable in the summers. At that time of the year, the feedlot owners prefer cattle with some Zebu blood, for they can tolerate the heat better.

The location of the feedlots is important. Feedlots should be near agricultural areas in order to obtain feed easily, as well as near the packing plants in the Phoenix area. They should also be near highway and rail routes so that feed and cattle can be economically and conveniently shipped. Thus feedlots need to be relatively close to the urban areas, but they should not be too close. A feedlot too near a city must deal with complaints about the flies and odor and pays higher taxes because of the inflated land values. A feedlot unfortunate enough to have a city grow out to it will be under great pressure to move its operations farther out (Map 7.4).

All feedlots are highly mechanized and automated and have a large food-processing factory associated with them. These factories are very modern and have large-capacity hay grinders, grain rolling mills equipped with steam plants, grain elevators, molasses blending machines, and other such equipment. The animals are fed a variety of food, such as alfalfa hay, milo, barley, cottonseed, vegetable by-products, molasses, and salt. In order to reduce the dust and provide quick energy to the animals, the cattle are fed

CATTLE SHIPPED INTO ARIZONA, 1978

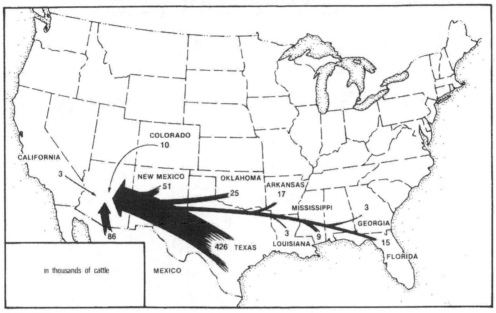

MAP 7.2
Source: Arizona Crop and Livestock Reporting Service, *Arizona Agricultural Statistics,* 1978.

CATTLE SHIPPED OUT OF ARIZONA, 1978

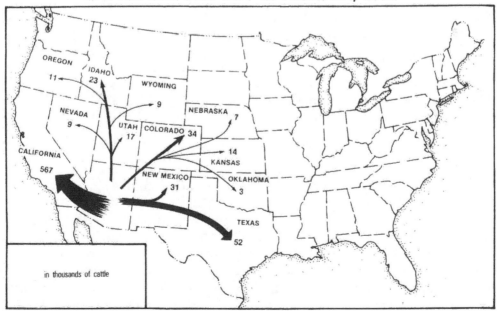

MAP 7.3
Source: Arizona Crop and Livestock Reporting Service, *Arizona Agricultural Statistics,* 1978.

FEEDLOTS IN ARIZONA, 1978

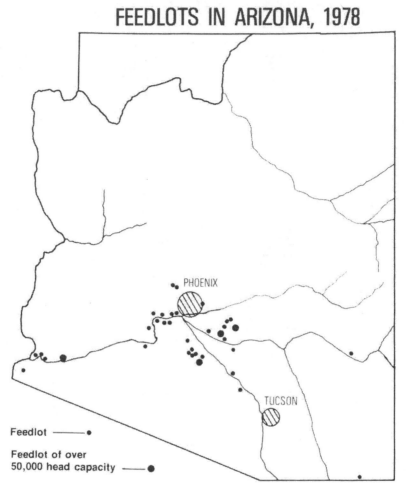

MAP 7.4
Source: Arizona Cattle Feeders Association, *Arizona Cattle Growers Outlook*, August 1978.

stabilized fat from already slaughtered animals. As the animals move into the feedlot, they are fed exacting diets, and that diet changes as they gain weight. One of the first things fed these animals is 3 lb. (1.4 kg.) of plastic pellets; these do not pass through the animal, so no further roughage is needed and there is less manure to remove. These pellets can also be used again after the animal is slaughtered. After the feed is mixed by the mill, trucks with hoppers dump the feed into troughs two to five times a day.

The dairy industry closely resembles the feedlot industry. The dairy farms are large (the average herd in late 1978 had 353 head), and the cattle are kept confined in small enclosures and fed processed feeds. Arizona dairies are extremely competitive and efficient. Arizona ranks second nationally in yearly average production, and the average milk cow in 1978 produced 15,936 lb. (7,235 kg.) of milk. Over 90 percent of the milk cows in the state are Holsteins, large black and white cows. Dairymen like the Holstein because it is a big milk producer. The cattle are fed very high-protein foods, and they are literally "burned out" after two years, three at the most. Then they are sold for slaughter, and the fact that they produce a large carcass is the second reason Holsteins are preferred by dairymen.

The location of the dairies is greatly influenced by the location of the market. Fresh milk, the major dairy product, is both bulky and perishable, so it is a distinct advantage for dairies to be near the consumers, as long as they are not so close as to cause problems (as with the feedlots). The principal market is Phoenix, and for that reason, most dairies and 86 percent of all dairy cattle are near that city. Central Arizona gets quite hot during the summer months, so milk production falls during that part of the year, but the demand also drops as the schools are not buying milk then.

Other Livestock

Besides cattle, other animals are of significance in Arizona's economy, particularly sheep — although sheep are not as important as they once were. The total number of sheep fluctuates from one year to the next (from 490,000 in 1977 to 483,000 in 1978), but the total number of breeding ewes in the state has stabilized at about 300,000 animals. About two-thirds of these Arizona sheep are owned by the Navajo. Although a great deal of government effort has been put into improving the quality of the Navajo sheep and in reducing the numbers of sheep on the overgrazed ranges, there are still problems in both of these areas. The government has also tended to emphasize raising sheep for meat production rather than for wool, and although this emphasis has helped the food supply, it has not helped the Navajo rug making.

Some non-Indian sheep are kept on farms, but a large number are moved back and forth between the mountains in summer and the farms in winter (Figure 7.18), a type of herding called transhumance. The sheep are not owned by the farmers, and the sheep owners must pay 12 cents per ewe per day for them to graze in the winter pastures, which are also the maternity wards and nurseries. The lambs are born in November, the shearing takes place in February and March, the young lambs are sold in April (when they weigh from 80 to 100 lb. [36 to 45 kg.]), and soon after that, the ewes return to their summer pastures in mountains. All the sheep now are trucked at least part way to the mountains.

There are many problems connected with this transhumance type of herding. One of these is labor; it is hard to find someone willing to spend entire summers herding sheep in the isolated mountains. Basque men have traditionally done this job, but it is common now to find that the herders come from the Andes. It is getting expensive to winter the sheep in alfalfa fields,

FIGURE 7.18. A herd of sheep being driven across a road in a ponderosa forest in northern Arizona.

and sheep must be watched closely when they first enter the fields or there will be deaths from bloat. Another problem in the winters is that sheep are killed by roving packs of dogs, mostly from the surrounding suburban homes. There is also less pastureland available than there used to be. These problems could lead to some modifications in the old herding system.

Hog and poultry farms are of little consquence to the state. Hog production is widely dispersed, and a number of communities have only a single producer. The only large concentration of hogs in the state is near Snowflake, where one of the world's largest hog farms is located. This one producer dominates the hog production in Arizona. The poultry industry specializes in the production of eggs, so it is located on the periphery of Phoenix, the major market.

MINING INDUSTRY

The mining industry, along with agriculture and cattle, prompted the early economic development of Arizona. This industry generates a great deal of wealth for the state, and in 1977 mineral production was set at $1.56

billion. The mining industry has not grown within recent years, and since the manufacturing and service sectors of the economy have boomed, the overall contribution of mining has declined. It is, however, critically important to dozens of small towns scattered across the state.

Mining and smelter towns are quite unique. Most mining towns are in mountainous areas, and the streets are not in the grid pattern found in most agricultural towns but run up and down the valleys and hills, following the contours (Figure 5.4). The level land in these mining towns is very valuable, and it is usually occupied by company buildings, businesses, churches, and schools. The homes, meanwhile, cling to the hillsides and are usually of poor quality and in bad states of repair. Nevertheless, these towns are quaint and interesting places to visit, and they attract many tourists.

Smelter towns, often within a few miles of the mining towns, are different. These are planned communities on more level land, such as Douglas, Clarkdale, and Hayden. These smelter towns seem to have a greater amount of permanency, but they, too, have poverty, and they lack the quaint character of the mining settlements.

The mining industry is not evenly scattered across the state (Map 7.5). The mineralized portions of the state are generally found in the Mountain District and in the Basin and Range Province. The Plateau section of the state has traditionally had little mining, but within recent years, uranium and coal mines have changed that picture. With only one or two exceptions, such as Jerome, the old mining districts of 1900 still exist as mining areas.

Early Development of Mining

The first miners in the state were the Indians, but they did only a little digging for such things as turquoise, coal, and salt. The Spanish were also miners in the area, but they did not prospect as did the early Americans. The Spanish "discovered" ore bodies mostly by asking the Indians for the locations of strange-looking rocks. Since Arizona was isolated, and relatively large ore bodies were being exploited much further south, the Spanish were little interested in anything but the richest of ores.

The Spanish used only very crude mining techniques, and they had no large, successful mines in the area. Traditionally, they would dig a shaft and follow a vein, and they would then crush and concentrate the ore at the mine site with an *arrastra* (Figure 7.19). This mining was done without explosives and usually with Indian help. The Spanish mined exclusively in the southern part of Arizona and even there, in only a very few areas. They did most of their mining near the Santa Cruz River, such as in the Patagonia area, and in the Santa Rita and Santa Catalina mountains. Because of Apache threats, the Spanish mining ended prior to the Gadsden Purchase.

The California gold rush of 1848–1849 greatly stimulated later gold exploration in Arizona. Many miners who had crossed the state and had not

MAJOR MINERAL AREAS OF ARIZONA

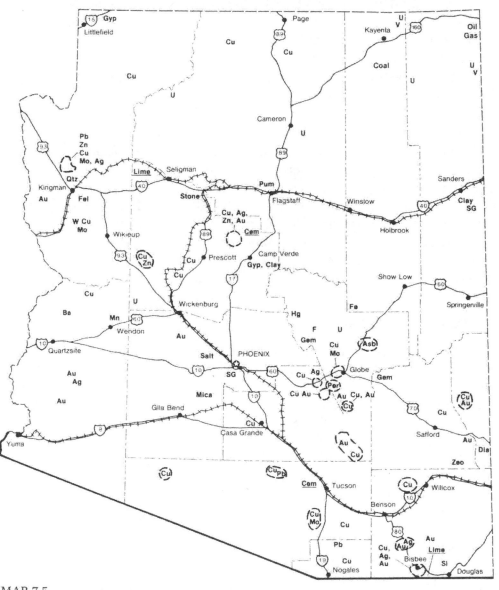

MAP 7.5
Source: U.S. Department of the Interior, Bureau of Mines, *Minerals in the Economy of Arizona* (Washington, D.C., 1978), pp. 6–7.

LEGEND

————	International boundary
– – – –	State boundary
— – – —	County boundary
✪	Capital
●	City
┣┼┼┼┫	Railroad
————	Road
▽	Interstate highway
▢	U.S. highway

Mineral Symbols

Ag	Silver ore
Asb	Asbestos
Au	Gold ore
Ba	Barite
Cem	Cement plant
Clay	Clay
Coal	Coal
Cu	Copper ore
Dia	Diatomite
F	Fluorspar
Fe	Iron ore
Fel	Feldspar
Gas	Natural gas
Gem	Gemstones
Gyp	Gypsum
Hg	Mercury
Lime	Lime plant
Mica	Mica
Mn	Manganese
Mo	Molybdenum
Oil	Petroleum, crude
Pb	Lead ore
Per	Perlite
Pum	Pumice
Qtz	Quartz
SG	Sand and Gravel
Salt	Salt
Si	Silica sands
Stone	Stone
U	Uranium ore
V	Vanadium ore
W	Tungsten ore
Zeo	Zeolite
Zn	Zinc ore
⌒	Concentration of mineral operations

```
0   10  20  30  40    miles
0   16  32      64    kilometers
```

FIGURE 7.19. An example of an *arrastra*. Many remains of these old *arrastras* are found in Arizona, since they were being built well into the twentieth century. Source: John Ross Browne, *Adventures in the Apache Country* (New York: Harper and Brothers, 1869).

struck it rich in California began to drift into Arizona looking for gold. This included both Americans and Mexicans, who were known mostly as Sonorans. Another stimulant to exploration in Arizona was the Gadsden Purchase, and several old Spanish mines were reopened during the early American period.

The Civil War largely ended mining operations since the Indian problems at that time caused most mines to close. Tradition has it that one mine, the Mowry Mine near Patagonia, continued to produce lead for the Confederate cause after most other mines had ended operations (Figure 7.20). Far to the west of the Apache threats, prospecting and mining for gold continued very successfully. Placer gold was found along the Colordo and lower Gila rivers and in the streams of the Bradshaw Mountains, such as Lynx Creek, Big Bug Creek, Hassayampa River, and others. Some gold mines were also opened at this time, such as the Vulture Mine near Wickenburg.

Gold remained significant after the Civil War, but at that time, silver pro-

FIGURE 7.20. A view of the Mowry Mine about 1869. Source: John Ross Browne, *Adventures in the Apache Country* (New York: Harper and Brothers, 1869).

duction in Arizona began to boom with the development of silver mines near Globe, Superior, Tombstone, and elsewhere. Large amounts of silver were produced, and Arizona was noted for its silver production. Mining at this time was of great overall importance to Arizona. In 1880, 21 percent of the labor force was working in the mines, and the only category to have more workers than mining employed people who were classified as "laborers." After every boom there comes a bust, and the one in silver came in 1893. Two things occurred then to cause the collapse of the silver industry: the repeal of the Sherman Silver Purchase Act and the exhaustion of the richer silver deposits.

Copper mining became important only after the railroads came into the area. In fact, the first railroad in Arizona was built to the mining district at Clifton-Morenci. Copper is basically an inexpensive ore that has to have cheap transport in order to be competitive in price. By 1888, copper production in Arizona had become very important, and the value of the copper mined was greater than all other metals combined. Many old silver mining districts continued in production because copper was often found concentrated below shallow silver deposits. Many large copper-producing districts developed—most notably Clifton-Morenci, Bisbee-Douglas, Globe-Miami, Ray-Superior, Ajo, Jerome—and since World War II many new districts have been developed, such as in the area south of Tucson and near the towns of Safford, San Manuel, Casa Grande, and Bagdad.

Some changes occurred in copper mining after 1900. There was a great demand for copper because of the spreading use of electricity, but the richest copper deposits were mostly depleted. However, Arizona had vast amounts of low-quality ores, and techniques for processing and handling this ore, such as the flotation process, were developed. These new techniques, along with the demand, led to a rapid growth of the copper industry in Arizona. By 1910 Arizona led the nation in the production of copper, and it has done so ever since.

Mining Today

Today Arizona accounts for over 60 percent of all copper mined in the United States (Figure 7.21). Copper consistently accounted for 85 to 90 percent of the value of all minerals produced in the state during the 1960s and through 1974, but that percentage began a gradual decline to 78 percent in 1977 because of a drop in copper prices, a growing interest in other ores, and the spectacular rise in the value of gold (Figure 7.22). Arizona is also a major producer of gold, silver, and molybdenum, ranking no less than fifth among the states in their production. Since a large portion of these metals—as well as others such as lead and zinc and more recently uranium—are recovered as by-products of copper mining operations, any curtailment in copper production will also affect these ores.

FIGURE 7.21

COPPER PRODUCTION IN THE UNITED STATES, 1977

State	Production (short tons)	Percent
Arizona	932,005	61.4
Utah	193,700	12.8
New Mexico	167,100	11.0
Montana	89,000	5.8
Nevada	66,850	4.4
Michigan	43,400	2.9
Other	25,983	1.7

Source: American Bureau of Metal Statistics

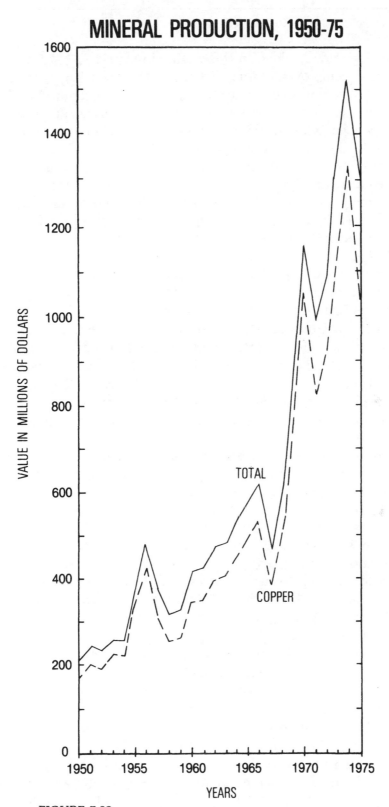

FIGURE 7.22
Source: Bureau of Mines, *Minerals Yearbook*, various years.

Many Arizonans are dependent on the mining industry. In 1977, 21,800 people were directly employed by the mines, and these workers were paid $381 million. Unionization occurred in the early twentieth century, and today the miners, as well as people working in milling, smelting, and refining operations, earn a good living. According to the Arizona Mining Association, an industry group, one out of every eight jobs in the state is dependent on mining. This figure includes mine and smelter employees, as well as those who provide the tools, services, and equipment to the mines and the miners. It is hard to estimate the total value added to the state's economy by the mines, but it is a very high figure. The total value of the minerals mined in Arizona in 1977 totaled $1,557,111,000.

The mining industry is thus still very important to the state's economy. It does not consume a great deal of water, it generates much wealth, and it uses only about one-quarter of 1 percent of the land in Arizona. In one respect, however, Arizona's mineral wealth is being exploited. Compared to the rest of the nation, Arizona has a "colonial" economy; relatively cheap raw materials (copper mostly) are exported, and relatively expensive manufactured products imported. The situation is changing, however, as manufacturing in the state increases, and more and more of Arizona's raw materials are being processed through to a finished product in the state. For example, in 1980 there are two rod plants in the state taking the finished copper and making it into wire.

Arizona is rich in minerals, but several factors determine what is mined and what is left unexploited. Economics is the dominant factor. Often it is cheaper to import foreign ores, which means that local mining suffers. Many foreign ore deposits are very rich, while the rich deposits of ore have already been mined in Arizona. Also, in order to sell on the world market, foreign governments will subsidize their home industries. The only way American corporations can compete is to have tariffs or limits placed on the foreign ores, or to stockpile American ores, but the U.S. government is reluctant to act in this case. Proximity to transportation is also a critical factor affecting the price. Also significant are the quantity of ore and the technologies to handle low-quality ores economically. Arizona has large ore deposits that are unexploited, but in the future, when cheap foreign ores are exhausted, they will prove to be very valuable.

The known copper reserves in Arizona are at present double what they were in the early 1900s. This is due to exploration and the utilization of the low-grade ores. That utilization requires massive amounts of capital and a developed technology, such as large-scale excavating and hauling machinery, improved drilling and blasting techniques, and the development of modern recovery methods such as flotation. These techniques are best utilized in an open-pit mine, and most of Arizona's great copper mining enterprises are based on such technology and the use of open-pit mines (Figure 7.23). There are, of course, still some rich deposits in the state, such as at the Superior Mine where miners sometimes hit ore of 10 or 20 percent copper, but today, a high percentage is rare and not the norm upon which the copper industry in Arizona is based. The average copper ore mined in the state is 0.6 percent copper. The average mine has to remove 2 tons (1.8 t.) of waste rock to mine 1 ton (.907 t.) of ore that will produce only 10 lb. (4.54 kg.) of copper, and copper sold for about one dollar a pound in early 1980. There are two basic types of copper ore, oxide ore and sulfide ore. The oxide ore is found above the other, and it has been weathered so it is often found in brilliant shades of blue and green. The techniques for handling the two types of ore are quite different.

Most of Arizona's copper comes from open-pit mines (Figure 7.24); of the 25 mines in the state, 20 are worked as open pits (Map 7.6 and Figure 7.25). The underground mines are very large and modern operations, such as the Superior and San Manuel mines and the Miami East Mine near Globe. The fourth underground mine is the Lakeshore Mine on the northern end of the Papago Reservation. This is a very modern and new mine, but it has experienced cost overruns and has only recently been reopened. The Old Reliable Mine uses neither the underground nor open-pit mining techniques. Rather, ore-bearing rock is periodically blasted, and a weak acid is poured on the mountain; as the acid percolates through, it dissolves the copper and then the acid, with the dissolved copper, is collected and processed. This type of mining is termed in-place or in-situ leaching. In-place leaching is also used at many of the large mines and at some of the closed mines (such as the mines at Bisbee), and in the future, this technique will be an important part of mining in Arizona.

In the second half of the 1970s, the copper industry in Arizona went through its worst economic crisis since the depression of the early 1930s. In 1974 copper was selling for 85 cents a pound, but in the next few years it was sold for a much lower price–down to 60 cents a pound, which is less than the cost of production at many U.S. mines and smelters. The reason for the price fall was a glut of copper on the world market. Third World copper-producing nations, such as Chile, Peru, Zambia, and Zaire, had to pay sharply higher prices for oil after 1973 and had incurred huge debts, so they expanded their state-owned copper industries in order to increase their foreign exchange. The result was greatly reduced copper prices and hard times in the mining towns in Arizona, because the Arizona mines could not compete with the government-subsidized ore. It was not until early 1979 that the copper industry in Arizona began to show signs of returning to prosperity; as the price of copper rose, mines reopened, and the mining companies expanded their operations.

Gold and silver ores are also extracted in Arizona, and they are of considerable value. Whereas there once were very large silver and gold mines in the state, there are none today. However, there is considerable gold and silver ore, and with the great increase in the price of gold, there has been a

FIGURE 7.23. A view of the Bagdad Mine, an open-pit type of mine.

FIGURE 7.24. A very large truck of the type used to remove ore from the open-pit mines. Rather than trucks, a few of the mines use railroads to remove the ore.

renewed interest in gold mining. Today, all of the gold and silver produced in the state comes as a by-product of other mining operations, particularly copper mining (Figure 7.26). The gold and silver, and various other metals and impurities, form a sludge at the bottom of the tank in the electrolytic refinery, and this sludge is put into barrels for shipment out of the state for processing. It is often said that the value of the gold and silver produced by the copper mines pays for the smelting process, and this is largely true, par-

COPPER MINING

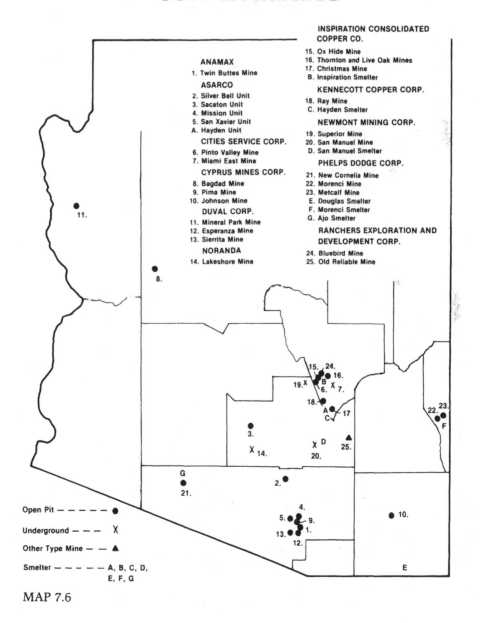

ANAMAX
1. Twin Buttes Mine

ASARCO
2. Silver Bell Unit
3. Sacaton Unit
4. Mission Unit
5. San Xavier Unit
A. Hayden Unit

CITIES SERVICE CORP.
6. Pinto Valley Mine
7. Miami East Mine

CYPRUS MINES CORP.
8. Bagdad Mine
9. Pima Mine
10. Johnson Mine

DUVAL CORP.
11. Mineral Park Mine
12. Esperanza Mine
13. Sierrita Mine

NORANDA
14. Lakeshore Mine

INSPIRATION CONSOLIDATED COPPER CO.
15. Ox Hide Mine
16. Thornton and Live Oak Mines
17. Christmas Mine
B. Inspiration Smelter

KENNECOTT COPPER CORP.
18. Ray Mine
C. Hayden Smelter

NEWMONT MINING CORP.
19. Superior Mine
20. San Manuel Mine
D. San Manuel Smelter

PHELPS DODGE CORP.
21. New Cornelia Mine
22. Morenci Mine
23. Metcalf Mine
E. Douglas Smelter
F. Morenci Smelter
G. Ajo Smelter

RANCHERS EXPLORATION AND DEVELOPMENT CORP.
24. Bluebird Mine
25. Old Reliable Mine

Open Pit — — — — — ●
Underground — — — X
Other Type Mine — — ▲
Smelter — — — — — A, B, C, D, E, F, G

MAP 7.6

FIGURE 7.25

MAJOR ARIZONA COPPER-PRODUCING MINES, 1978

Mine	Pounds Recoverable Copper[*]	Pounds Recoverable Molybdenum
1. Morenci	270,934,080	–
2. San Manuel	234,783,000	3,452,101
3. Twin Buttes	205,907,000	3,130,000
4. Sierrita	199,154,769	16,338,357
5. Ray	171,217,513	632,758
6. Pinto Valley	145,596,000	450,000
7. Bagdad	137,296,652	2,577,425
8. Metcalf	108,779,990	–
9. New Cornelia	84,309,533	–
10. Mission	79,485,782	375,239
11. Superior	77,429,000	–
12. Inspiration	74,210,213	61,507
13. Sacaton	46,084,526	
14. Silver Bell	42,531,344	133,776
15. Mineral Park	30,351,809	4,512,456
16. San Xavier	24,898,978	–

[*]Total copper in the ore, but not all of it can be recovered.

Source: Arizona Department of Mineral Resources.

ticularly since the price of gold has risen so high.

The story of the lead and zinc industry is somewhat similar to that of the copper industry. There is a considerable amount of lead and zinc ore in Arizona, but because of foreign competition, little of either is mined today. The peak of lead and zinc production was in 1949, when a combined total of over 100,000 tons (90,700 t.) were produced. Yavapai County is the leading lead and zinc area, and the mines there are staying in operation because of trace amounts of gold and silver in the ore. The primary zinc producing area of the United States is in the Missouri-Oklahoma-Arkansas area, and all the zinc mined in the Bagdad area is sent to Bartlesville, Oklahoma, for processing. If and when the prices of these ores rise appreciably on the world market, Arizona could once again be a major producer of lead and zinc.

The second major ore produced in Arizona, by value, is molybdenum (it is commonly called molly). It is a little known ore and does not have romanticism associated with it, as do copper, gold, and silver. There are no separate molybdenum mines in the state, and all of it is produced as a by-product of copper. Its major use is as an alloy, and it is shipped to the major steel producing centers in the United States and abroad, particularly Japan.

Various other metals are also found in the state, such as manganese, mercury, tungsten, and others, but the production of these ores is small and

FIGURE 7.26

SOME METALS PRODUCED AS A BY-PRODUCT OF COPPER, 1977

Metal	Production
Molybdenum	34,574,477 Lbs.
Silver	6,815,040 Tr. Oz.
Gold	94,000 Tr. Oz.

Source: United States Bureau of Mines.

sporadic, depending on demand and price. The asbestos industry is an example of this sporadic mining. At one time, there were many asbestos mines in the state, mostly in the area to the north and east of Globe. Arizona traditionally leads the nation in asbestos production, but most of the demand now is for short-fiber asbestos to be used as filters in the chemical industry, and most of that type of asbestos is produced in large Canadian mines. The potential for a large asbestos production is present in Arizona, but at present asbestos mining is only a small operation.

Uranium is another important potential resource in Arizona. There was a boom in uranium exploration and mining after World War II, but there was little interest in the mineral through the 1960s and the early 1970s. Since the late 1970s there has been an increased interest in uranium, and various large companies have staked out potentially rich areas. Most of the early mining of uranium was from the sedimentary rock formations on the Colorado Plateau, particularly on the Navajo Reservation, but much of the exploration since the late 1970s has been south of the Plateau. There is a relatively high uranium content in the ore at the Twin Buttes Mine south of Tucson, a copper mine, and since early 1980 the mine has been concentrating the uranium and shipping it to Oklahoma for processing. This company plans to produce up to 140,000 lb. (63,560 kg.) of yellowcake uranium annually.

There are also fossil fuels in the state. There is a small amount of oil and gas extraction in northern Apache County, but it is of little overall significance to the state. There is, however, quite a bit of coal. The single largest deposit of coal is in the Black Mesa field, but there are scattered smaller deposits in eastern Arizona (Map 7.7).

The Indians were the first to mine the Black Mesa coal, and between 1300 and 1600 they mined over 100,000 tons (90,700 t.). Between 1925 and 1960, about 10,000 tons (9,070 t.) were mined annually for Indian consumption and use in surrounding towns, but then production fell as other fuel sources, such as gas, were more widely used. In the mid-1960s, the Peabody Coal Company acquired leases from the Navajo and the Hopi, which allowed the mining of about 120 million tons (109 million t.) of coal over a

COAL IN ARIZONA

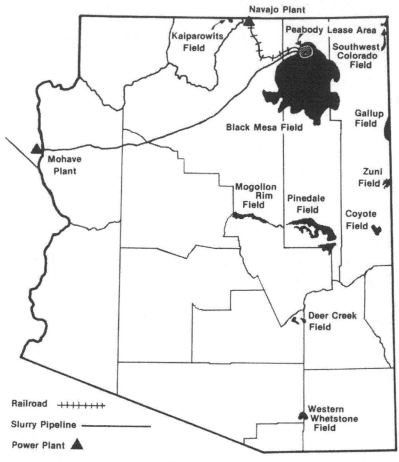

MAP 7.7
Source: Arizona Bureau of Mines, *Coal, Oil, Natural Gas, Helium, and Uranium in Arizona*, Bulletin 1982, 1970.

35-year period. The initial contract calls for the Indians to be paid a royalty of 20 cents per ton for coal used on the reservation and 25 cents per ton for coal used elsewhere. There is also a graduated scale for increasing royalties as the cost of coal rises, and by 1980 they were receiving an average of 54 cents per ton for coal mined on the reservation. This royalty arrangement is not a high one, and the Indians would like to renegotiate the contract.

The Peabody Coal Company has been actively mining coal on the Black Mesa since 1970. The coal in that area is quite good, having a fairly high heat content and low sulfur and ash contents. It is fairly thick (4 to 30 ft. [1.2 to 9.1 m.]) and relatively near the surface, so it is being strip mined at the rate

FIGURE 7.27. A large dragline stripping overburden at the Kayenta Mine on Black Mesa. The main boom is considerably longer than a football field, and the bucket has a capacity of 90 cubic yards (68.8 cu. m.). For scale, note the bulldozer on the left. Use of such large machines is common in the West and offers economies of scale. Smaller machines are used in the pit to load the coal on trucks, and later the land is releveled.

of about 400 acres (162 ha.) each year (Figure 7.27). There are two mines 5 mi. (8 km.) apart—the Black Mesa Mine is on the joint-use land, and the Kayenta Mine is on Navajo land. A determined effort is being made to level the overburden and to establish vegetation on it. The Indians benefit in that most of the laborers are Indians, but unfortunately, the Hopi have strong feelings against such activity, so almost all of the workers are Navajo.

The mined coal is moved to two very large and modern power plants (Map 7.7). The Mohave power plant is on the Colorado River in Nevada, and the coal is shipped there as slurry from the Black Mesa Mine through an underground pipeline that is 18 in. (45.7 cm.) thick, which was completed in 1972. The powdered coal is mixed with water from deep wells, made into a slurry, and two and a half days after being put into the pipeline, it emerges in southern Nevada. The other plant receiving coal from the Black Mesa is the Navajo power plant, a few miles from Page, and the coal is shipped there from the Kayenta Mine—by conveyor belt for the first 5 mi. (8 km.) and then on the specially built, electrically powered Black Mesa and Lake Powell Railroad for 78 mi. (125 km.). These two power plants need vast amounts of water for cooling purposes, hence their location along the Colorado River – the Navajo power plant, for example, has the right to annually use over 34,00 acre-feet of water. There are also several other large, modern coal-fired power plants in the Four Corners area, but they receive their coal from elsewhere, particularly from New Mexico.

Another important mined product, ranking third by value and first by tonnage, is sand and gravel, which is taken mostly from the riverbeds to be used in building and highway construction. It is a cheap product that cannot stand shipment costs, so it must be mined near where it is used. Most construction is near Phoenix, so Maricopa County leads the state in sand and gravel production, and Pima County is second.

LUMBER INDUSTRY

Although Arizona is known as a desert state, it has a considerable lumber industry since about 26 percent of the state is forested (Map 2.7). Compared to other western states, such as Oregon and Washington, Arizona produces little lumber, but what is produced is of major importance to the state and to the local economies.

Arizona has many native species of trees, but few of those are used commercially, and 99 percent of the potential lumber is in the softwood category. Arizona's forest can be grouped into two general classifications, commercial or sawtimber forest and noncommercial or woodland forest. Arizona has 3,180,000 acres (1,287,900 ha.) of commercial forest land, and much of it is virgin forest, a rarity in the United States. These forests are capable of producing commercial timber crops, such as sawlogs, pulpwood, poles, and posts. Approximately 90 percent of the commercial forests contain trees of sawtimber size (trees 12 in. [30.5 cm.] or larger in diameter at 4.5 ft. [1.4 m.] above the ground), and the rest contain pole timber (trees 5 to 12 in. [12.7 to 30.5 cm.] in diameter). These forests primarily consist of four species suitable for sawtimber or pole timber on a commercial basis – Douglas fir, white fir, Engelmann spruce, and ponderosa pine. The first three species, together with a few other softwoods (especially corkbark fir and aspen) and a hardwood (Gambel oak), compose only 12 percent of the sawtimber. Douglas fir, the most important of these species, makes up about 7.3 percent of the sawtimber, begins growing at about 7,500 ft. (2,286 m.) in elevation, and white fir and spruce are found in the subalpine zone above 8,500 ft. (2,591 m.). Aspen, which requires a very high and moist environment, is found in patches throughout the forests, but it comprises less than 1 percent of the sawtimber available.

The ponderosa pine is far and away the most important tree crop in the state; it constitutes over 87 percent of the sawtimber, and currently ponderosa pine accounts for about that percentage of the total cut. In

Arizona, this pine reaches its best development between 7,000 and 8,000 ft. (2,134 and 2,438 m.), and it is usually found in pure stands, a trait favored by the lumbermen. The main stand is found along the Mogollon Rim, extending from south and west of Williams all the way eastward into New Mexico. This is the largest single stand of ponderosa pine in the world, over 300 mi. (483 km.) broad and from 20 to 60 mi. (32 to 96 km.) wide. North and south of this stand, ponderosa can be found growing in isolated spots at the higher elevations.

The yields from the ponderosa forests are low in comparison to the yields from forests in more humid areas, averaging between 5,000 and 15,000 board feet per acre. The ponderosa, however, is a "near-desert" pine capable of growing on sites too dry for other sawtimber species, and it has other advantages. It produces an excellent, all-around wood, it usually grows in pure stands, and it easily replenishes itself from one generation to the next. The chance of fire is greater in this type of forest because of its dry character, but once started, a fire is more easily controlled because the crown cover is more open than those of the heavier stands of spruce and fir.

Noncommercial forests in Arizona occupy 16 million acres (6.5 million ha.) and can be subdivided into three general groups—mesquite, chaparral, and piñon-juniper. Mesquite grows in several areas and is only of local importance, mostly as firewood. Chaparral is commonly found at 4,000 to 6,000 ft. (1,219.2 to 1,828.8 m.) and is of little use. Piñon-juniper is of only little more value than the other two, and it is mainly found just below the ponderosa zone (5,000 to 6,500 ft. [1,524 to 1,981 m.]). These forests do provide some cover and support for wildlife and grazing for cattle, but today the three groups are considered only big weeds that need to be eliminated. Extensive programs are being carried out to eradicate these noncommercial forests and to have grass in their stead, because grass is good for the cattle, increases the runoff for cities and agriculture, and helps to control silting.

The control of Arizona's forests is in the hands of the federal government. Approximately 66 percent of the forests are located in national forests, 29 percent are on Indian reservations (particularly the Navajo and Apache reservations), 1 percent is owned by the state, and only about 4 percent is privately owned. The U.S. Forest Service, a bureau of the Department of Agriculture, has jurisdiction over the national forests, but it does not actually harvest and process the trees, as those operations are handled by private companies who buy the trees through competitive bidding. Trees are considered a crop and are treated as such by the forest service; they are planted, thinned, and protected from fire, disease, and insects. Thus, the forests are handled for the good of all through forest management techniques.

Most of the national forests in Arizona have exploitable amounts of timber (Map 7.8). Only Coronado National Forest in southern Arizona and Prescott National Forest are unable to support large lumber industries. Sitgreaves

ARIZONA LUMBER INDUSTRY

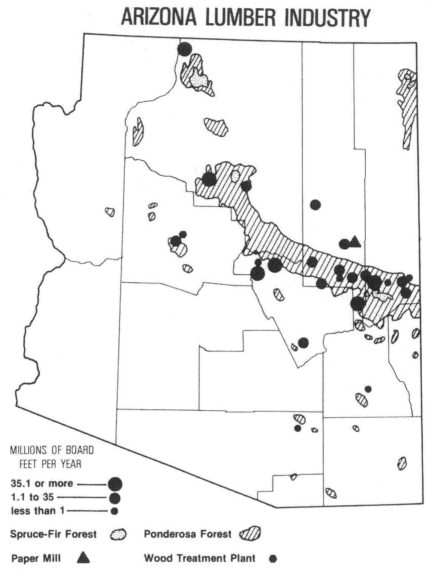

MILLIONS OF BOARD
FEET PER YEAR

35.1 or more ———— ●
1.1 to 35 ———— ●
less than 1 ———— ●

Spruce-Fir Forest ⬭   Ponderosa Forest ⬭

Paper Mill ▲   Wood Treatment Plant ●

MAP 7.8
Source: Arizona Land Department, *Arizona Land Marks, Primary Wood Industries of Arizona*, vol. 8, bk. 5, 1978.

National Forest, the smallest in Arizona but also the most heavily timbered, has an extensive forest industry. The Coconino National Forest near Flagstaff is near the main line of the Santa Fe Railroad, so it was the first forest to be extensively exploited. The section of the Kaibab National Forest north of the Grand Canyon also supports a large logging operation. Since it is so isolated on the Strip, the first large sawmill to use its timber was not built until 1952. Prior to that time, the logs had to be hauled to mills at Flagstaff and elsewhere, but now they need to be hauled only 60 mi. (96.5 km.) to Fredonia.

Logging is also actively carried out on the Indian reservations. When logging initially began on the Indian reservations in the late 1800s, the proceeds from the timber went to the federal government, but that situation changed in 1936 with the passage of the General Forest Regulation, which stated that the Indians were to use their forest to help support their populations. Most of the logging, however, continued to be done by outsiders, as the Indians slowly learned to manage and exploit their forests. McNary is a lumber town on the Fort Apache Indian Reservation, and the mill was completed in 1920 to exploit the Indians' timber. But the Indians have now become quite adept at forestry and now have their own sawmills, and beginning in 1970 the McNary sawmill couldn't use Indian timber and had to ship in logs from the national forests. The Fort Apache Indian Reservation is an example of how the Indians have successfully exploited their forests, because it has over 700,000 acres (283,500 ha.) of commercial forest and, with little help from outsiders, annually produces about 73 million board feet. Tribal lumber production increases tribal income, employment, and pride.

Arizona's forests have been exploited since humans first entered the state. The Indians used wood for construction and tools and as firewood, but probably the only way they truly affected Arizona's forests was through controlled burns of the undergrowth to open up the forest environment. The Spanish likewise had only a slight impact on the forests, since they confined their activities to the southernmost part of Arizona. The early Americans in the mining era had a great impact on local areas, and many scrub forests near the mines were decimated for the making of charcoal. The first year of recorded lumber production was 1869, when 1.2 million board feet were produced (an impressive figure at first glance, but today, it is considerably less than a single day's output). One of the first recorded sawmills in the state was brought overland from Utah by Mormons who erected the mill south of Flagstaff in 1878, at a site since known as Sawmill Springs. The first extensive exploitation of Arizona's forests began when the Atchison, Topeka and Santa Fe Railroad extended its line through the timbered section of Arizona near Flagstaff. That railraod mostly crossed desert areas, and thus it had a great need for crossties. The first lumber mill to produce crossties for the railroad was established as early as 1881, and success of that mill attracted many others. As population in the state grew, the demand for lumber increased. The great demand for lumber was by Anglos from the humid East who moved into the area; they had a tradition of building houses of sawed timber, and the local mills tried to meet their demand. Timber production grew throughout the twentieth century. By 1935, over 100 million board feet were produced, by 1956 that figure had increased to 300 million board feet, and in 1977 the state produced 392 million board feet with a wholesale value of $104 million.

The lumber industry is still important to the state's economy. In 1976, there were 4,000 people employed in the "lumber and wood products" industry. Since then, employment in that sector has soared, and by late 1978, it stood at 6,400 with an annual payroll of over $56 million. Most of the lumber industry is concentrated in three counties—Coconino, Navajo, and Apache. It is important to all three, but it is critically important to Apache County because in the early 1970s half of all employment there was in the forest industries.

Most of the forest products are sold outside the state. Traditionally, 90 percent of the products were sold elsewhere, but with the growing demand from Phoenix and Tucson, that figure has dropped to about 75 percent. In the early 1970s, the major markets were in the western states, but some timber was sent to the Midwest and some even to such distant states as Maine and Florida. Ponderosa pine is in great demand because it has a warm appearance, is easily worked, and has a uniform texture. The lumber industry, like the mining industry, illustrates a colonial type of economy: Arizona's lumber is shipped out of the state, often to be processed or manufactured into a product that is shipped back into the state for consumption. Arizona should encourage the wood-products industry to transform local wood into the highest-value product possible, by manufacturing charcoal, furniture, paneling, cabinetwork, and the like. Another possible industry would be the production of glue-laminated lumber products of fir and spruce capable of supporting great weight for use in construction. All of these manufacturing ventures would help the local economies by generating employment and taxes.

Although there are many sawmills in the state, there is only one large paper mill. The first such mill was established at Flagstaff in 1954, and that small business is still in operation and now converts wastepaper into paper towels. The importance of this mill is that it proved that ponderosa pine makes a quality paper. The largest paper mill in Arizona, completed in 1961, was built 13 mi. (21 km.) west of Snowflake by the Southwest Forest Industries. The mill receives its raw material from many sources; in all, the company receives sawdust and chips from 36 mills, most of which are in Arizona, but some are in the surrounding states. Much of the raw material comes by rail, but the local sawmills send it by truck. This mill now annually produces 258,000 tons (234,006 t.) of kraft paper and 154,000 tons (139,678 t.) of newsprint. The kraft paper is sent to the mill's 38 container-board plants, only one of which is in Arizona—in Glendale—and about 70 percent of the newsprint is sent to southern California. Most of the remaining newsprint is used in Arizona, but a small portion is sold in neighboring states. This paper mill shows what can be done; today it is an important industry that is based on a product that was formerly burned or thrown away.

Through careful planning, the mill has had a minimal effect on the local environment, and it is a major employer and generator of wealth in the Snowflake area.

Arizona's forest industry has a great potential for growth, since manufacturing techniques could be developed or expanded in the state and new industries could be started. The new industries could include plywood manufacturing, naval stores, and secondary manufacturing such as furniture, architectural woodwork, and cabinetwork. Pole processing is another possible industry. Currently there is only one commercial wood pressure-treating and preserving plant in the state. Completed in 1961 it is on land leased from the Yavapai Indians, and it is owned by Southwest Forest Industries. Prior to the completion of this plant, all wooden power and telephone poles had to be imported, but today, this plant produces crossties, power poles, fence posts, mine timbers, and so forth. The state government should encourage private industry to use Arizona's forests to their fullest, while maintaining a sustained yield.

MANUFACTURING

Arizona traditionally has had a colonial economy – raw goods are shipped out of the state and manufactured goods are imported – but that picture is rapidly changing. An indication of the development of manufacturing is the growth of manufacturing employment: in 1939 only 7,996 Arizonans were employed in manufacturing, but by 1967 that figure had grown to 76,800, and in mid-1978 it stood at 119,600. Thus Arizona is rapidly moving in the direction of having a more industrialized economy.

The common method of measuring manufacturing output is to determine the value added to a product by a manufacturing process. The idea of value added is important because it indicates the total amount of money generated by the manufacturing process, and it indicates, in a general way, the amount of labor needed. The value added, for example, varies greatly between a pound of steel made into shovels, and that same pound of steel made into cameras.

The total value added for all manufacturing in Arizona in 1977 was estimated at $3.38 billion. This was more than double the value of the minerals mined in the state and about three times the value of crop and livestock production. Since agriculture and mining have grown only slightly over the last quarter century, the great expansion of Arizona's economy in this period has largely been the result of its expanding industrial base.

Food processing was the major manufacturing employer prior to 1950. This industry handles the large amount of food grown in the state that has to be processed before it can be exported. It is a lower-wage industry and produces little wealth. Food processing is still important in Arizona, but today it

FIGURE 7.28

ARIZONA EMPLOYMENT IN MANUFACTURING, 1976

Industry	Value Added by Manufacture (in thousands)	Number of Employees
Machinery, Except Electric	$ 614.8	10,913
Electric, and Electronic Equipment	469.2	19,660
Primary Metal Industries	375.0	7,462
Transportation	355.1	15,761*
Food and Kindred Products	197.9	6,588
Printing and Publishing	164.4	7,433
Instruments and Related Products	136.0	4,665
Chemicals and Allied Products	126.6	2,315
Stone, Clay, Glass	111.3	3,249
Fabricated Metal Products	95.3	4,560
Apparel and Other Textile Products	74.6	6,098
Lumber and Wood Products	62.5	3,965
Paper and Allied Products	35.7	1,018
Furniture and Fixtures	26.4	1,210
Rubber and Misc. Plastic Products	5.1	1,416
Misc. Manufacturing Industries	12.8	2,500
TOTAL	$ 2,862.7	98,813

*interpolated from figures given.

Source: U.S. Bureau of the Census, Annual Survey of Manufacturers, 1976; U.S. Bureau of the Census, County Business Patterns, 1976: Arizona.

employs less than 7 percent of all manufacturing employees.

Four major sectors dominate manufacturing in Arizona: (1) machinery, (2) electronics, (3) primary metals, and (4) transportation. These four accounted for almost 63 percent of all manufacturing in the state in 1976, and the other 37 percent was distributed among the other industrial sectors (Figure 7.28). Three of the four, however, actually involve electronics and aerospace, but that fact is hidden by the classifications assigned by the federal government. Machinery is the largest sector in terms of value added, and it employs about 11,000 workers. The largest category within the machinery sector is office and computing equipment, which is mostly related to electronics. The second sector in significance is electric and electronic equipment, employing 19,660 people. The fourth sector, transportation, includes such things as aircraft and parts, guided missiles, and space vehicles and parts—industries obviously related to the electronics industry. There is even another sector related to electronics, "instruments and related products."

The electronics and aircraft industries clearly dominate the manufacturing portion of the state's economy. They employ great numbers of people, but even more importantly, these industries pay high wages and are non-polluting operations. They are what is termed light industry and are very desirable. The single largest private employer in the state is Motorola, with approximately 16,000 workers statewide, about 6 percent of all manufacturing employees. Other similar employers include AiResearch, Goodyear Aerospace, Sperry Flight Systems, Honeywell Information Systems, Hughes Aircraft, General Electric, Siemens Corporation, and several smaller firms. A great amount of technical expertise is needed in these industries, expertise not previously needed in Arizona, so many of these employees are relative newcomers to the state.

The third major manufacturing sector is "primary metal industries." This sector of the economy is closely tied to the copper industry and currently employs well over 7,000 workers. Another important sector is "printing and publishing," which is mostly the newspaper industry. As mentioned, food processing is still significant, and so is the apparel and textile industry, but these two sectors employ mostly cheap and unskilled labor and therefore have a smaller impact on the state's economy. The other industries are of lesser significance.

One noteworthy aspect of manufacturing in Arizona is that it is concentrated in the metropolitan areas. Maricopa County, with 55 percent of the state's population, has 74.3 percent of the manufacturing employment, and Pima County, the only other area in the state with a large population concentration, has 19.1 percent. Therefore, the Phoenix and Tucson areas have 93.4 percent of all manufacturing employment, a tremendous concentration. It is, thus, understandable why so many outsiders who move to Arizona choose to settle in one of these two areas – the money and the jobs are there.

The greatest significance of manufacturing in Arizona is that many of the manufactured goods are shipped out of state. By value, the bulk of Arizona's exports are electronics and machinery, and these exports totaled $442.6 million in 1976. Almost 33 percent of all the machinery produced and 29 percent of all the electronics were exported. These are very high figures, and they mean that a great deal of money flowed into the state.

There are various reasons for the rapid growth of Arizona's manufacturing industry. One reason is Arizona's life-style. It was easy to attract skilled laborers to Arizona from the Northeast, because the climate is pleasant and the environment is open and accessible. Once there was a large, skilled labor force available, even more industry tended to move to Arizona. Air-conditioning really made it all possible, for the summers in Phoenix and Tucson are hot, yet air-conditioning means that life can continue and industrial production can remain high through the summer months. There are also examples of some companies moving to southern Arizona because their president, or some other officer within the company, enjoys spending the winters there. Another factor is the relative cost of energy; electrical energy has traditionally been cheaper in Arizona.

Manufacturing is one of the principal employers in the state, directly employing about one in eight workers. Most of the manufacturing is concerned with electronics and aerospace, is characterized by high technology and high wages, and is in the forefront of development. It will continue to grow, and all of Arizona will benefit.

TOURISM

Tourism has been important in Arizona's economy for a long time, but it is dependent upon good transportation facilities—before the era of the railroads, few tourists came to Arizona. The railroads opened up Arizona, and by the 1880s, many tourists did more than just cross the territory, they stopped to visit or to enjoy Arizona's reportedly healthful climate. A cowboy-and-western mystique had developed by the 1890s, and dude ranches became scattered across Arizona because some ranchers found it more profitable to cater to tourists than to cattle. The Grand Canyon has been attracting tourists since 1885 when a road was completed to it from Flagstaff. A stagecoach was taking tourists to the Grand Canyon by 1892, and a boom in tourism began in 1901 when the Santa Fe extended a line to the South Rim. Tourism began to "hit its stride" at that time, and it has grown ever since. By the 1920s, a good highway system was developing in Arizona, and that greatly stimulated automobile traffic. There was little tourism during the years of World War II, but since then, tourism has been important in the state's economy.

Tourism had a $3.4 billion impact on the state in 1978. Tourism is a labor-intensive industry that produces many jobs, directly employing 71,614 people in 1978 and indirectly employing another 119,959 for a total of 191,573 Arizonans dependent on tourism. Tourism ranked second to manufacturing among Arizona's industries, far ahead of mining and agriculture in economic significance. Another advantage to this industry is that it is spread over the state and greatly benefits the economies of many smaller towns. It even helps Arizonans visiting their own state, as there are more facilities and amenities scattered over the state to serve the out-of-state tourists. Tourists from the Phoenix and Tucson areas visiting out-of-the-way spots in Arizona also contribute to the economies of the small towns in the state.

The tourists now travel to Arizona mostly by car or by air; bus and train travel, once popular, are now of little consequence for Arizona's tourists. The spending and travel habits of the airline and auto tourists are quite different, even though they spend about the same amount of money in the state

FIGURE 7.29

EXPENDITURES OF HIGHWAY AND AIRLINE TOURISTS IN ARIZONA, 1977

Category	Highway Tourists		Airline Tourists	
	Total	Percent	Total	Percent
Lodging	$ 289,938,820	22.2	$ 217,464,133	18.9
Food	264,562,123	20.3	179,601,340	15.7
Transportation	225,103,799	17.2	349,482,359	30.4
Other	524,837,219	40.3	399,107,346	34.8
TOTAL	$1,304,441,961		$1,145,655,178	

Source: M.E. Bond and Bill McDonald, "Tourism: An Arizona Growth Industry," Arizona Business, June/July, 1978.

(Figure 7.29). The tourists who enter Arizona by air are mostly from distant states. The greatest number are from California, but there are a fair number from other populous industrial states, such as Illinois, Texas, Pennsylvania, Ohio, and New York. The vast majority enter Arizona through the two largest cities, particularly Phoenix, and more than 65 percent have one of those two cities as their primary destination (they may not go elsewhere). Business-related travel dominates, and most business people travel alone. About 30 percent of the air travelers are professional people, and 12 percent are retired. Visiting friends and relatives and recreation are, respectively, the second and third reasons air travelers go to Arizona. The typical airline traveler stays only three days, but during the winter and spring the length of stay noticeably increases.

The typical highway tourist in Arizona is from a nearby state. California leads the list, followed by Texas, New Mexico, and Colorado, and the Midwestern states come next. In the summer months, the typical highway traveler usually crosses northern Arizona, is one of a party of three (one a child), is traveling between southern California cities and the Midwest, and only considers Arizona a place to cross as quickly as possible. This has always been the case, though more people are now considering Arizona to be their destination or are spending longer periods of time there. In the winter season, the typical highway travelers are generally older, travel to southern and central Arizona, and spend a longer time in the state. There are also many interstate travelers, usually urban residents who travel by car for one to three days. These tourists spend considerably less than the out-of-state tourists, but they do make a significant contribution to the economies of the smaller towns.

Tourists spend money for a variety of services, needs, and desires. The average tourist spends roughly the same amounts for food, lodging, and

FIGURE 7.30

TOTAL TOURISM AND TRAVEL EXPENDITURES IN ARIZONA, 1977

Category	Expenditure	Percent
Lodging	$ 560,085,470	19.4
Food	581,163,210	20.2
Transportation	689,229,574	23.9
Other	1,054,158,858	36.5

Source: M.E. Bond and Bill McDonald, "Tourism: An Arizona Growth Industry,"
Arizona Business, June/July, 1978.

transportation (Figure 7.30), although the typical air traveler spends considerably more for transportation. The tourists also spend a considerable amount of money for entertainment, souvenirs, and general retail purchases. The air travelers spend 95 percent of their money in the two metropolitan areas, and the highway travelers spend only half there and the other half throughout the state. Thus any gasoline shortage has a great impact on the outlying regions and relatively little influence on the metropolitan areas.

There are also many foreign travelers in the state. Because Mexico is so close, many of its citizens travel to Arizona, and in the year ending March 1978, it is estimated that Mexicans spent $315,324,000 in Arizona. Most of the Mexican tourists were interested in shopping, which particularly benefited the border towns and Tucson; they were actually more shoppers than tourists. As an indication of this very casual nature of their visits, the largest category of expenditure was for goods bought in department stores, and the third category was for purchases in grocery stores (Figure 7.31). Canadian visitors are another significant group of foreign tourists. Most are far wealthier than the average Mexican visitors, and typically the Canadians spend much of the winter in southern Arizona. Perhaps not surprisingly, the foreigners, mostly Mexicans, accounted for about 25 percent of all bus travelers.

Arizona derives obvious advantages from the tourist industry, such as employment and the money spent, but the problems caused by this industry are also evident. Crowding and congestion during the summer months in northern Arizona and the winter months in the south are a major concern. Recreational sites are often quite crowded because of out-of-state visitors, and during the winter season in southern Arizona the traffic congestion and shopping delays caused by the winter residents, or "snowbirds," make many Arizonans question the economic worth of the winter tourist industry. The long-term winter visitors in particular cause problems, because they drive up the price of temporary housing, and services, such as sewage and

FIGURE 7.31

EXPENDITURE BY MEXICAN VISITORS TO ARIZONA

FOR THE YEAR ENDING MARCH 1978

Category	Total Spent
Lodging	$ 9,707,000
Restaurants	9,210,000
Groceries	45,257,000
Transportation	12,788,000
Department Stores	137,131,000
Medical or Health	6,627,000
Business Expenditures	77,859,000
All Other	16,745,000
TOTAL	$315,324,000

Source: Nat de Gennaro and Robert J. Ritchey, "Mexican Visitors to Arizona Identified as Valued Market," Arizona Review, August–September, 1978.

hospitalization, that are used effectively for only a few months each year must be added for them. The tourists also add to the water and air pollution problems. Jobs connected with the tourist industry do not pay well, and although that situation is improving, the growth of total personal income is below that for other industries such as manufacturing and mining. Many of the jobs are also seasonal, so that a person working in the tourist industry is usually low paid and out of work part of the year. Nevertheless, tourism remains an important and growing industry.

GENERAL ECONOMY

The economy of Arizona is healthy. It is quite diversified—with mining, manufacturing, agriculture, tourism, and various other sectors—but Arizona's economy is closely tied to the overall national economy, and any fluctuations in the latter will quickly have a direct bearing on the state.

All industries and jobs can theoretically be classified into two broad groups, those that are "basic," and those that are service oriented, or "nonbasic." Basic industries are those industries that bring money into the state and upon which a state or region is dependent, and they can be divided into three categories. First, basic industry can be extractive. This type of industry takes from nature, such as ranching, mining, lumbering, or agriculture, and can be renewable or, as in the case of mining, nonrenewable. Second, manufacturing is also a basic industry, for it produces much wealth from the sale of locally made products out of state. The third category is basic revenue from other sources, such as the money spent by tourists and retirees living in Arizona, and from federal government employees working in the state. Any state must have basic industries so that purchases of out-of-state goods can be made, and the more basic industries a state has, the "richer" it is. Arizona is fortunate to have a great deal of basic industry and a great variety of it.

Service or nonbasic industries are those industries that provide service and support to the citizens but are very dependent on the basic sector of the economy to bring money into a region. Examples of service industries are restaurants, laundries, retail stores, banks, and construction firms – all are dependent on basic industries. Should the basic industry in a one-industry town, such as a mining or a military town, suffer or close, the service sector of that town's economy is affected accordingly. In an industrialized nation or state the service industry employs a high percentage of the population. The industry that is particularly dependent on growth is construction, and as a result of growth there was a great boom in the construction industry in the 1970s.

In June of 1978, 935,400 Arizonans were employed. The largest category of employees were those working for governments (Figure 7.32), and the federal employees in that sector, 36,700 people, were very important in that they spent federal money in Arizona. This would definitely be a "basic" part of the economy. Arizona has a larger percentage of government employees

FIGURE 7.32

EMPLOYMENT IN ARIZONA, JUNE 1978

Occupation	Total People Employed	Percent
Manufacturing	119,200	14.3
Mining and Quarrying	20,000	2.4
Construction	62,700	7.5
Transportation	20,000	2.4
Communications and Public Utilities	24,300	2.9
Wholesale Trade	40,600	4.9
Retail Trade	162,700	19.5
Finance, Insurance, and Real Estate	47,300	5.7
Services and Miscellaneous	155,600	18.7
Government	180,100	21.6

Source: Valley National Bank, Arizona Statistical Review, 1978.

than the average state, possibly because of the number of national parks, forests, and monuments in the state, but there are also many military personnel in Arizona.

There are several large military bases in the state, and they all have a great impact on local economies (Figure 7.33). The one large army base in the state is Fort Huachuca, which is the site of the army's Electronic Proving Grounds and the headquarters for STRATCOM (the army's Strategic Communications Command). Fort Huachuca is very important in the overall economy of southeastern Arizona. There are approximately 17,000 military and their dependents stationed on this base and over 4,700 civilian employees. There are also many thousands of others dependent on the base in an indirect way, and it is far and away the largest employer in Cochise County, which has a total population of only 79,500. If the base were to close, it would cause economic havoc in that entire corner of the state.

Other important military posts were opened in Arizona just prior to the time the United States entered World War II with the construction of air training bases. Williams Field and Luke Field were opened near Phoenix, and Davis-Monthan Field, near Tucson. These are still large, active air force bases and are important in the local economy. An army proving ground and a marine air station have also been built near Yuma. Thus, the military still plays a large and direct role in the economy of Arizona, and there are also many retired military personnel in the state who contribute greatly to the overall economy.

The second and third largest employment sectors, on the basis of wages and salaries paid, are "retail trade," such as store, restaurant, and filling station employees and owners, and "services and miscellaneous," such as employment in hotels and other lodgings and business and medical services.

FIGURE 7.33

ECONOMIC IMPACT OF ARIZONA MILITARY BASES ON LOCAL ECONOMIES

Base	Location Area	Impact on Local Economy
Davis-Monthan	Tucson	$140,653,587 (1977)
Luke (includes Gila Bend)	Phoenix	107,117,861 (1977)
Williams	Phoenix	62,187,746 (1978)
Ft. Huachuca	Sierra Vista	187,600,000 (1979)
Yuma Proving Ground	Yuma	35,004,722 (1978)
Marine Corps Air Station	Yuma	42,624,198 (1978)
Navajo Depot Activity	Flagstaff	2,066,033 (1978)

Source: Impact Statements and Communication from Military Bases.

These are obviously service industries, as are several employment sectors (construction, transportation, communications and public utilities, whole-sale trade, and finance, insurance, and real estate). The large number of people classified as "service employees" indicates a mature economy.

Arizona's economy is viable and strong. More and more people are moving into the state, and this population growth stimulates the economy. Most of the new migrants to Arizona are in their young and most productive years, and many seek employment in manufacturing. Thus, while employment in agriculture, ranching, and mining has been rather stable, employment in manufacturing has grown by leaps and bounds.

8

CONCLUSION

The physical and human phases of the geography of Arizona are diverse, and few generalizations can be drawn about the state as a whole. It is, rather, a state of vivid contrasts—from one region to the next and from one time period to those that follow. Thus, rather than generalizing, it is easier to draw comparisons and contrasts.

Arizona can be divided into three major physical regions: the Plateau, the Basin and Range, and the transition zone—the Mountain District. There is tremendous variety among these regions in such things as land forms, vegetation, resources, and human utilization, but within each district there is some uniformity, and generalizations can be made. The Plateau region is high, open, and cut by deep crevasses. The highest portions are forested, but most of it is desert or near desert. Human use of it is varied. The Indians, particularly the Navajo and the Hopi, occupy much of it and use it for grazing. Most other residents of the Plateau region live along the east-west line of communication, and there is only one significant town, Flagstaff. The Basin and Range, meanwhile, is also open and flat, but it is divided by long, narrow mountain ranges and is almost exclusively desert. It is noted for its dense populations, rich agricultural oases, and mining districts and for its hot summers now made bearable by evaporative coolers and air-conditioning. It is difficult to categorize the Mountain District. It is a tremendous jumble of mountains, some of which are rich in minerals and mining towns, and others are not, some of it is high enough to be forested, but much of it is desert, and so it goes. Variety is the rule in the mountains.

The history of peoples in Arizona is also a story of variety. The Indians have lived in Arizona for a very long time, but they have made little impact on the environment. Long before the Spanish arrived, the practice of agriculture had spread to Arizona, and with it the Indians evolved organized

317

and relatively highly developed societies that had collapsed or were in stages of decline when the Spanish arrived. Although the Spanish along the upper Rio Grande extended some influence into Arizona, the only area of Latin control in present-day Arizona was along the Santa Cruz Wash from Tucson southward. Thus, with the Gadsden Purchase, the United States acquired about 1,000 potential Mexican-American citizens in what was to be Arizona.

American control did not lead to rapid changes in Arizona. It was not until the 1880s that the "hostile" Indians had been controlled and effective transportation extended to Arizona. The Americans then began ranching, mining, and agricultural operations on a broad scale, and began affecting the environment in a negative way.

Arizona now has a polyglot society. It has more full-blooded Indians than any other state, and those Indians still have strong tribal ties. There is also a strong Mexican-American element in the population as well as many other minority and ethnic groups. Those Arizonans classified as "white" make up 72 percent of the population. They have come from almost everywhere to settle Arizona, but the largest number came by way of California and the Midwest.

The outsiders' image of Arizona is one of cowboys, and that image is strong even within the state. Western clothes are a popular item, and western bars are well attended; but most cowboys in the state are of the Saturday night variety. It may be a false image, but it is a strong one nevertheless, even among the Arizonans, and it has greatly affected their character.

Arizona is an urban state. The majority of Arizonans, about 76 percent, live in either the Phoenix or the Tucson metropolitan areas. It is in these two places that the jobs are located and money can be earned, so these areas are growing rapidly. The growth in these cities only encourages more growth, while the small towns of Arizona remain small and often tied to one industry.

Arizona's population is rapidly growing, and in terms of percentage it now ranks either first or second in the nation. As industry, jobs, and people move to the state, many question the value of all the growth. It does provide more jobs, but the drawbacks are great and include increased pollution of air and water, traffic jams, and increased pressure on recreational areas. Perhaps Arizona can do only a little toward limiting its growth, but it certainly can do more toward controlling, guiding, and preparing for it.

The economy of Arizona is strong, but that was not always the case. Arizona traditionally has had a boom to bust history, and that process has occurred repeatedly in the agricultural, mining, and ranching sectors of the economy. The situation has been changing, especially with the development of manufacturing, and that change indicates a maturing of the economy.

Arizona's economic base is also slowly changing. In the past, it was chic to say the economy rested on the "five Cs" – cattle, copper, cotton, citrus, and climate. That phrase covered the major industries, but it left out others, such as the various agricultural and mineral products and lumbering. The five Cs still remain, but today other industries have to be added, including construction – a factor of growth – computers and component manufacturing, and tourism (the last unfortunately does not start with a "C"). Changes are also coming in the relative importance of the various industries. Agriculture and mining are declining in relative significance, while the other sectors of the economy are growing. Yet agriculture and mining remain just as important to many small towns across the state as they were thirty or forty years ago.

Arizona faces many problems in the future, and one of the greatest of these is water. Much of Arizona is dependent on groundwater, and it is being removed far faster than natural recharge is occurring. Arizona will someday exhaust that water source, and that day probably will be in the not too distant future. Tucson is particularly vulnerable, because it is the largest city in the world to be wholly dependent on groundwater. The water laws are now being changed in order to better conserve and use the groundwater. Agriculture will be greatly affected in the long run; sooner if the new laws are for the good of all Arizonans, but if nothing is changed the farmers will run out of water anyway – it will just take longer. The dropping water table and the increasing cost of pumping are already causing the farmers to evaluate alternate crops and other possible ways to use less water.

Water will also be a problem in Phoenix. Although Phoenix now has plenty of groundwater and surface water, that picture will change. The groundwater is being depleted, and upstream from Phoenix the Apache Indians are beginning to tap the surface water as it flows along their reservations; the courts have decreed that the water is first reserved for the Indians' use, and someday the Phoenix metropolitan residents may have to buy water from the Indians. There are various other problems connected with water, but solutions to them are also available, and Arizonans are beginning to grapple with finding solutions to all the state's problems.

The future of Arizona is bright. Although many new residents are moving to the state, it is still sparsely populated and has room for growth. Arizona has many resources, not the least of which are its residents, both old and new.

SUGGESTED ADDITIONAL READING

INTRODUCTION

Arizona, University of, Faculty. 1972. *Arizona–Its People and Resources.* Tucson: University of Arizona Press.

Barnes, William C. 1960. *Arizona Place Names.* Tucson: University of Arizona Press.

de Roos, Robert. 1963. "Arizona: Booming Youngster of the West." *National Geographic* 123:299–343.

Meinig, Donald W. 1971. *Southwest: Three Peoples in Geographical Change.* New York: Oxford University Press.

_____ . 1972. "American Wests: Preface to a Geographical Interpretation." *Annals of the Association of American Geographers* 63:159–84.

PHYSICAL BACKGROUND

Benson, Lyman. 1969. *The Cacti of Arizona.* Tucson: University of Arizona Press.

Brown, Arthur A., and Davis, Kenneth P. 1973. *Forest Fire: Control and Use.* New York: McGraw-Hill Company.

Dunbier, Roger. 1968. *The Sonoran Desert: Its Geography, Economy, and People.* Tucson: University of Arizona Press.

Durrenberger, Robert. 1972. "The Colorado Plateau." *Annals of the Association of American Geographers* 63:211–36.

Green, Christine R., and Sellers, William D. 1964. *Arizona Climate.* Tucson: University of Arizona Press.

Hastings, James R., and Turner, Raymond M. 1965. *The Changing Mile: An Ecological Study of Vegetation Change with Time in the Lower Mile of an Arid and Semi-Arid Region.* Tucson: University of Arizona Press.

Humphrey, Robert R. 1970. *Arizona Range Grasses.* Tucson: University of Arizona Press.

Kangieser, Paul C. 1966. "Climates of the States: Arizona." In U.S. Department of Commerce, Environmental Science Services Administration, *Climatology of the United States,* nos. 60-62. Washington, D.C.: Government Printing Office.

Kearney, Thomas H.; Peebles Robert H.; et al. 1960. *Arizona Flora.* Berkeley: University of California Press.

Kozlowski, Theodore T., and Ahlgren, C. E., eds. 1974. *Fire and Ecosystems.* New York: Academic Press.

Lowe, Charles H. 1964. *Arizona's Natural Environment.* Tucson: University of Arizona Press.

Lowe, Charles H., and Brown, D. E. 1973. *The Natural Vegetation of Arizona.* Phoenix: Arizona Resources Information System.

Martin, S. Clark, and Turner, Raymond M. 1977. "Vegetation Change in the Sonoran Desert Region, Arizona and Sonora." *Journal of the Arizona Academy of Science* 12:59-69.

EARLY SETTLEMENT OF THE LAND

Bancroft, Hubert H. 1962. *History of Arizona and New Mexico.* Albuquerque, N.M.: Horn and Wallace.

Bingham, Edwin R. 1960. *The Fur Trade in the West, 1815-1846.* Boston: Heath.

Bishop, Morris. 1933. *The Odyssey of Cabeza de Vaca.* New York: Century Company.

Bolton, Herbert Eugene. 1932. *The Padre on Horseback.* San Francisco: Sonora Press.

———. 1936. *Rim of Christendom.* New York: Macmillan Company.

———. 1949. *Spanish Exploration in the Southwest, 1542-1706.* New York: Barnes and Noble.

Brandes, Raymond. 1960. *Frontier Military Posts of Arizona.* Globe, Ariz.: Dale Stuart King.

Chittenden, Hiram M. 1954. *The American Fur Trade of the Far West.* Stanford, Calif.: Academic Reprints.

Cleland, Robert Glass. 1950. *This Reckless Breed of Men: The Trappers and Fur Traders of the Southwest.* New York: Alfred A. Knopf.

Conkling, Roscoe P., and Conkling, Margaret B. 1947. *The Butterfield Overland Mail.* Glendale, Calif.: A. H. Clark Company.

Corle, Edwin. 1951. *The Gila: River of the Southwest.* New York: Holt, Rinehart and Winston.

Coues, Elliott, trans. 1900. *On the Trail of a Spanish Pioneer: The Diary and Itinerary of Francisco Garces.* 2 vols. New York: Francis P. Harper.

Cremony, John C. 1969. *Life Among the Apache.* Glorieta, N.M.: Rio Grande Press.

Day, Arthur Grove. 1964. *Coronado's Quest.* Berkeley: University of California Press.

Doyel, David E. 1979. "The Prehistoric Hohokam of the Arizona Desert." *Amer-*

ican Scientist 67:544–54.

Faulk, Odie B. 1970. *Arizona, A Short History.* Norman: University of Oklahoma Press.

Garber, Paul N. 1959. *The Gadsden Treaty.* Gloucester, Mass: Peter Smith.

Gladwin, Harold S.; Haury, Emil W.; et al. 1965. *Excavations at Snaketown.* Tucson: University of Arizona Press.

Goetzmann, William H. 1959. *Army Explorations in the American West, 1803–1863.* New Haven: Yale University Press.

Haines, Francis. 1971. *Horses in America.* New York: Thomas Y. Crowell.

Hallenbeck, Cleve. 1940. *Alvar Nuñez Cabeza de Vaca.* Glendale, Calif. Arthur H. Clark Company.

Hallenbeck, Cleve, ed. and trans. 1949. *The Journey of Fray Marcos de Niza.* Dallas, Tex: University Press.

Haury, Emil W. 1967. "The Hohokam: First Masters of the American Desert." *National Geographic* 131:670–95.

———. 1976. *The Hohokam Desert Farmers and Craftsmen.* Tucson: University of Arizona Press.

Kidder, Alfred Vincent. 1962. *An Introduction to the Study of Southwestern Archaeology.* New Haven: Yale University Press.

Kirk, Bryan. 1929. "Flood-Water Farming." *Geographical Review* 19:444–56.

McGregor, John C. 1965. *Southwestern Archaeology,* Urbana: University of Illinois Press.

Marshall, James. 1945. *Santa Fe: The Railroad That Built an Empire.* New York: Random House.

Martin, Paul S., and Plog, Fred. 1973. *The Archaeology of Arizona.* Garden City, N.J.: American Museum of Natural History.

Mattison, Ray H. 1946. "Early Spanish and Mexican Settlement in Arizona." *New Mexico Historical Review* 21:273–327.

Myrick, David F. 1975. *Railroads of Arizona.* Berkeley, Calif.: Howell-North Books.

Paré, Madeline F., and Fireman, Bert. 1965. *Arizona Pageant: A Short History of the 48th State.* Phoenix: Arizona Historical Foundation.

Pattie, James Ohio. 1962. *The Personal Narrative of James Ohio Pattie.* Philadelphia: Lippincott.

Polzer, Charles W. 1968. "Legends of Lost Missions and Mines." *Smoke Signal,* no. 18:170–83.

Proctor, Gil. 1956. *Tucson, Tubac, Tumacacori, Tohell.* Tucson: Arizona Silhouettes.

Singletary, Otis A. 1960. *The Mexican War.* Chicago: University of Chicago Press.

Smith, Fay J.; Kessell, John L.; and Fox, Francis J. 1966. *Father Kino in Arizona.* Phoenix: Arizona Historical Foundation.

Spicer, Edward H. 1962. *Cycles of Conquest: The Impact of Spain, Mexico, and the United States on the Indians of the Southwest, 1533–1960.* Tucson: University of Arizona Press.

Stewart, Kenneth M. 1965. "The Southwest." In Robert F. Spencer and Jesse D. Jennings, eds., *The Native Americans,* pp. 283–336. New York: Harper and Row.

Terrell, John Upton. 1968. *Estevanico the Black.* Los Angeles: Westernlore Press.

Trimble, Marshall. 1977. *Arizona: A Panoramic History of a Frontier State.* New

York: Doubleday and Company.

Turney, Omar A. 1929. *Prehistoric Irrigation in Arizona.* Phoenix, Ariz.: O. A. Turney.

Utley, Robert M. 1967. *Frontiersmen in Blue: The United States Army and the Indian, 1848-1865.* New York: Macmillan.

Walker, Henry P., and Bufkin, Don. 1979. *Historical Atlas of Arizona.* Norman: University of Oklahoma Press.

Wilson, Neill C., and Taylor, Frank J. 1952. *Southern Pacific.* New York: McGraw Hill.

Wormington, H. M. 1947. *Prehistoric Indians of the Southwest.* Denver, Colo.: Denver Museum of Natural History.

Wyllys, Rufus K. 1950. *Arizona: The History of a Frontier State.* Phoenix, Ariz.: Hobson and Herr.

POPULATION AND ETHNIC GROUPS

Allen, Edward Jones. 1936. *The Second United Order Among the Mormons.* New York: Columbia University Press.

Arrington, Leonard J. 1958. *Great Basin Kingdom: An Economic History of the Latter-Day Saints.* Cambridge: Harvard University Press.

Bailey, Paul D. 1971. *City in the Sun: The Japanese Concentration Camp at Poston, Arizona.* Los Angeles: Westernlore Press.

Breed, Jack. 1948. "Land of the Havasupai." *National Geographic* 93:655-74.

Brodie, Fawn M. 1946. *No Man Knows My History: The Life of Joseph Smith the Mormon Prophet.* New York: Alfred A. Knopf.

Brooks, Juanita. 1961. *John Doyle Lee: Zealot, Pioneer, Builder, Scapegoat.* Glendale, Calif.: A. H. Clark Company.

_____ . 1962. *The Mountain Meadows Massacre.* Norman: University of Oklahoma Press.

Brown, Malcolm, and Cassmore, Orin. 1939. *Migratory Cotton Pickers in Arizona.* Washington, D.C.: Government Printing Office.

Carroll, John M., ed. 1971. *The Black Military Experience in the American West.* New York: Liveright.

Castetter, Edward F., and Bell, William H. 1942. *Pima and Papago Indian Agriculture.* Albuquerque: University of New Mexico Press.

Craig, Richard B. 1971. *The Bracero Program.* Austin: University of Texas Press.

Durham, Philip, and Jones, Everett L. 1965. *The Negro Cowboys.* New York: Dodd, Mead and Company.

Francaviglia, Richard V. 1970. "Mormon Landscape." *Proceedings of the Association of American Geographers* 2:59-61.

_____ . 1978. *The Mormon Landscape.* New York: AMS Press.

Golder, Frank A., ed. 1928. *The March of the Mormon Battalion.* New York: Century Company.

Goodwin, Grenville. 1969. *The Social Organization of the Western Apache.* Tucson: University of Arizona Press.

Hill, Willard W. 1938. *The Agricultural and Hunting Methods of the Navajo Indians.* New Haven: Yale University Press.

Hoover, J. W. 1929. "The Indian Country of Southern Arizona." *Geographical Review* 19:38–60.

———. 1930. "Tusayan: The Hopi Indian Country of Arizona." *Geographical Review* 20:425–44.

———. 1937. "Navajo Land Problems." *Economic Geography* 13:281–300.

Joseph, Alice; Spicer, R.; and Chesky, J. 1949. *The Desert People: A Study of the Papago Indians.* Chicago: University of Chicago Press.

Katz, William Loren. 1971. *The Black West.* Garden City, N.Y.: Doubleday and Company.

Kitano, Harry H. L. 1969. *Japanese Americans.* Englewood Cliffs, N.J.: Prentice Hall.

Kluckhohn, Clyde, and Leighton, D. 1962. *The Navajo.* Garden City, N.Y.: Natural History Library.

Lamb, Blaine. 1977. "Jews in Early Phoenix." *Journal of Arizona History* 18:299–318.

McClintock, James H. 1921. *Mormon Settlement in Arizona: A Record of Peaceful Conquest of the Desert.* Phoenix, Ariz.: Manufacturing Stationers, Inc.

Meinig, Donald W., 1965. "The Mormon Culture Region." *Annals of the Association of American Geographers* 55:191–220.

Myer, Dillon S. 1971. *Uprooted Americans: The Japanese Americans and the War Relocation Authority During World War II.* Tucson: University of Arizona Press.

O'Kane, Walter C. 1953. *The Hopis.* Norman: University of Oklahoma Press.

Papanikolas, Helen Z., ed. 1976. *The Peoples of Utah.* Salt Lake City: Utah State Historical Society.

Peterson, Charles S. 1973. *Take Up Your Mission: Mormon Colonizing Along the Little Colorado River, 1870–1900.* Tucson: University of Arizona Press.

Prago, Albert. 1973. *Strangers in Their Own Land: A History of Mexican-Americans.* New York: Four Winds Press.

Riley, Carroll L. 1972. "Blacks in the Early Southwest." *Ethnohistory* 19:247–60.

Sanders, Thomas G. 1979. "Mexico: After the Revolution That Failed." *Focus* (American Geographical Society) 29, no. 5:1–12.

Savage, W. Sherman. 1976. *Blacks in the West.* Westport, Conn.: Greenwood Press.

Servin, Manuel P. 1970. *The Mexican-Americans: An Awakening Minority.* Beverly Hills, Calif.: Glencoe Press.

Spier, Leslie. 1928. "Havasupai Ethnography." *Anthropology Papers of the American Museum of Natural History* 29, part 3.

Tipton, Gary P. 1977. "Men Out of China: Origins of the Chinese Colony in Phoenix." *Journal of Arizona History* 18:341–56.

Tyler, Daniel. 1969. *A Concise History of the Mormon Battalion in the Mexican War 1846–1847.* Glorieta, N.M.: Rio Grande Press.

Underhill, Ruth M. 1967. *The Navajos.* Norman: University of Oklahoma Press.

U.S., Bureau of Indian Affairs. 1978. *Information Profiles of Indian Reservations in Arizona, Nevada, and Utah.* Washington, D.C.: Government Printing Office.

Van Willigen, John. 1970. "Contemporary Pima House Construction Practices." *Kiva* 36:1–10.

Waters, Frank. 1963. *Book of the Hopi.* New York: Ballantine Books.

Weaver, Thomas, ed. 1974. *Indians of Arizona: A Contemporary Perspective.* Tucson: University of Arizona Press.

Weber, David J., ed. 1973. *Foreigners in Their Native Land: Historical Roots of the Mexican Americans.* Albuquerque: University of New Mexico Press.

Young, Karl E. 1968. *Ordeal in Mexico.* Salt Lake City, Utah: Deseret Book Company.

URBANIZATION AND URBAN GROWTH

Gibson, Lay J. 1975. "Tucson's Evolving Commercial Base, 1893–1914: A Map Analysis." *Historical Geography Newsletter,* 5, no. 2:10–17.

Gibson, Lay J., and Reeves, Richard. 1970. "Functional Bases of Small Towns: A Study of Arizona Settlements." *Arizona Review* 19:19–26.

Jackovics, Theodore W., and Saarinen, Thomas F. 1978. "The Sense of Place: Student Impressions of Tucson and Phoenix." *Arizona Review* 27, no. 4:1–12.

Jeffery, David. 1977. "Arizona's Suburbs of the Sun." *National Geographic* 152: 486–517.

McLaughlin, Herb, and McLaughlin, Dorothy, eds. 1970. *Phoenix, 1870–1970.* Phoenix: Arizona Photographic Associates.

Mawn, Geoffrey P. 1977. "Promoters, Speculators, and the Selection of the Phoenix Townsite." *Arizona and the West* 19:207–24.

Peterson, Gilbert J. 1970. "The Townsite Is Now Secure: Tucson Incorporates, 1871." *Journal of Arizona History* 11:151–74.

Sargent, Charles S. 1975. "Towns of the Salt River Valley, 1870–1930." *Historical Geography Newsletter* 5, no. 2:1–9.

———. 1978. "Salt River Valley Urban Growth: Past-Present-Future." Scottsdale, Ariz.: SRVisions.

Wehmeier, Eckhard. 1975. *Die Bewässerungsoase: Phoenix, Arizona.* Stuttgarter Geographische Studien, ed. Wolfgang Meckelein and Christoph Borcherdt, vol. 89. Stuttgart: Geography Institute of the University of Stuttgart, 1975.

LAND AND WATER

Arizona, University of. 1977. *Arizona Water: The Management of Scarcity.* Phoenix: Arizona Academy.

Barr, James L., and Pingry, David E. 1977. "The Central Arizona Project." *Arizona Review* 26, no. 4:1–49.

Carstenson, Vernon R., ed. 1963. *The Public Lands: Studies in the History of the Public Domain.* Madison: University of Wisconsin Press.

Gay, Francois J. 1977. "La Competition pour l'utilisation de l'eau: besoins urbains & besoins ruraux en Arizona." *Cahiers geographiques de Rouen,* no. 6:41–73.

Greever, William S. 1954. *Arid Domain: The Santa Fe Railroad and Its Western Land Grant.* Stanford: Stanford University Press.

Harshbarger, J. W., et al. 1966. *Arizona Water.* Geological Survey Water-Supply Paper 1648. Washington, D.C.: Government Printing Office.

Henderson, Patrick C. 1966. "The Public Domain in Arizona, 1863–1891." Ph.D. dissertation, University of New Mexico.

Johnson, Richard. 1977. *The Central Arizona Project, 1918–1968.* Tucson: University of Arizona Press.

Kelso, Maurice M.; Martin, William E.; and Mack, Lawrence E. 1973. *Water Supplies and Economic Growth in an Arid Environment: An Arizona Case Study.* Tucson:

University of Arizona Press.

Linser, C. Laurence. 1977. "Alternative Futures: Arizona State Water Plan, Phase II." Arizona Water Commission.

Mann, Dean E. 1963. *The Politics of Water in Arizona*. Tucson: University of Arizona Press.

Mattison, Ray H. 1967. "The Tangled Web: The Controversy over the Tumacacori and Baca Land Grants." *Journal of Arizona History* 8:71–90.

Mosk, Sanford Alexander. 1944. *Land Tenure Problems in the Santa Fe Railroad Grant Area*. Berkeley: University of California Press.

Peterson, Dean F., and Crawford, A. Berry, eds. 1978. *Values and Choices in the Development of the Colorado River Basin*. Tucson: University of Arizona Press.

Powell, Donald M. 1960. *The Peralta Grant: James Addison Reavis and the Barony of Arizona*. Norman: University of Oklahoma Press.

Robbins, Roy M. 1976. *Our Landed Heritage: The Public Domain, 1776–1970*. Lincoln: University of Nebraska Press.

Templer, Otis W. 1978. "Texas Surface Water Law." *Historical Geography* 8:11–20.

Warne, William E. 1973. *The Bureau of Reclamation*. New York: Praeger Publishing Company.

ARIZONA'S EVOLVING ECONOMIC BASE

Arizona Bureau of Mines. 1961. *Gold Placers and Placering in Arizona*. Bulletin no. 168. Tucson: University of Arizona Press.

Arizona Crop and Livestock Reporting Service. 1979. "Arizona Agricultural Statistics, 1978." Bulletin S-14.

Bond, M. E., and McDonald, Bill. 1978. "Tourism: An Arizona Growth Industry." *Arizona Business* 25, no. 6:3–12.

Cleland, Robert Glass. 1952. *A History of Phelps Dodge, 1834–1950*. New York: Alfred A. Knopf.

Hill, James S.; Hillman, Jimmye S.; and Henderson, P. L., 1965. "Some Economic Aspects of the Arizona Citrus Industry." Agricultural Experiment Station, University of Arizona, Technical Bulletin 168.

Joralemon, Ira B. 1973. *Copper*. Berkeley, Calif.: Howell-North Books.

Jordan, Terry G. 1972. "The Origin and Distribution of Open-Range Cattle Ranching." *Social Science Quarterly* 53:105–21.

Kent, Kate Peck. 1957. *The Cultivation and Weaving of Cotton in the Prehistoric Southwestern United States*. Philadelphia: American Philosophical Society.

Line, Francis R. 1950. "Arizona Sheep Trek." *National Geographic* 97:457–78.

McGowan, Joseph C. 1961. *History of Extra-Long Staple Cottons*. El Paso, Tex.: Hill Printing Company.

Mayes, Horace M. 1974. "Cropland Atlas of Arizona." Arizona Crop and Livestock Reporting Service.

Pierce, H. W., and Wilt, J. C. 1970. "Coal, Oil, Natural Gas, Helium, and Uranium in Arizona." Arizona Bureau of Mines, Bulletin no. 182.

Smith, Bertie Webster. 1967. *The World's Great Copper Mines*. London: Hutchinson and Company.

Tatsch, J. H. 1975. *Gold Deposits: Origin, Evolution, and Present Characteristics*. Sudbury, Mass.: Tatsch Associates.

Wagoner, Jay J. 1952. *History of the Cattle Industry in Southern Arizona, 1540–1940*. Social Science Bulletin no. 20. Tucson: University of Arizona.

Young, Otis E. 1967. *How They Dug the Gold*. Tucson: Arizona Pioneer's Historical Society.

INDEX